Introduction to
Graph Theory
Solutions Manual

Introduction to
Graph Theory
Solutions Manual

Koh Khee Meng
National University of Singapore, Singapore

Dong Fengming
Tay Eng Guan
Nanyang Technological University, Singapore

World Scientific

NEW JERSEY · LONDON · SINGAPORE · BEIJING · SHANGHAI · HONG KONG · TAIPEI · CHENNAI

Published by

World Scientific Publishing Co. Pte. Ltd.

5 Toh Tuck Link, Singapore 596224

USA office: 27 Warren Street, Suite 401-402, Hackensack, NJ 07601

UK office: 57 Shelton Street, Covent Garden, London WC2H 9HE

British Library Cataloguing-in-Publication Data
A catalogue record for this book is available from the British Library.

INTRODUCTION TO GRAPH THEORY
Solutions Manual

ISBN-13 978-981-277-175-9 (pbk)
ISBN-10 981-277-175-1 (pbk)

Printed in Singapore.

Preface

Discrete Mathematics is a branch of mathematics dealing with finite or countable processes and elements. Graph theory is an area of Discrete Mathematics which studies configurations (called graphs) consisting of a set of nodes (called vertices) interconnecting by lines (called edges). From humble beginnings and almost recreational type problems, Graph Theory has found its calling in the modern world of complex systems and especially of the computer. Graph Theory and its applications can be found not only in other branches of mathematics, but also in scientific disciplines such as engineering, computer science, operational research, management sciences and the life sciences. Since computers require discrete formulation of problems, Graph Theory has become an essential and powerful tool for engineers and applied scientists, in particular, in the area of designing and analyzing algorithms for various problems which range from designing the itineraries for a shipping company to sequencing the human genome in life sciences.

Graph Theory shows its versatility in the most surprising of areas. Recently, the connectivity of the World Wide Web and the number of links needed to move from one webpage to another has been remarkably modeled with graphs, thus opening the real world internet connectivity to more rigorous studies. These studies form part of research in the phenomena of the property of a 'small world' even in huge systems such as the aforementioned internet and global human relationships (in the so-called 'Six Degrees of Separation').

This book is intended as a companion to our earlier book Introduction to Graph Theory (World Scientific, 2006). Here, we present worked solutions to all the exercise problems in the earlier book. Such a collection of solutions is perhaps the first of its kind. We believe that the student who has worked on the problems himself will find the solutions presented here useful as a check and as a model for rigorous mathematical writing. For ease of reference, each chapter begins with a recapitulation of some of the important concepts and/or formulae from the earlier book.

We would like to thank Prof. G.L. Chia, Ms Goh Chee Ying, Dr Jin Xian'an, Dr Ng Kah Loon, Prof. Y.H. Peng, Dr Roger Poh, Ms Ren Haizhen, Mr Soh Chin Ann, Dr Tan Ban Pin, Dr Tay Tiong Seng and Dr K.L. Teo for reading the draft and for checking through the solutions - any mistakes that remain are ours alone.

Koh Khee Meng
Dong Fengming
Tay Eng Guan
April 2007

Notation

$\mathbb{N} = \{1, 2, 3, \cdots\}$

$|S| = $ the number of elements in the finite set S

$\binom{n}{r} = $ the number of r-element subsets of an n-element set $= \frac{n!}{r!(n-r)!}$

$B \setminus A = \{x \in B \mid x \notin A\}$, where A and B are sets

$\bigcup_{i \in I} S_i = \{x \mid x \in S_i \text{ for some } i \in I\}$, where S_i is a set for each $i \in I$

$(\Rightarrow)$ proof of the implication "if P then Q" in the statement "P if and only if Q"

$(\Leftarrow)$ proof of the implication "if Q then P" in the statement "P if and only if Q"

[Necessity] proof of the implication "if P then Q" in the statement "P if and only if Q"

[Sufficiency] proof of the implication "if Q then P" in the statement "P if and only if Q"

In what follows, G and H are multigraphs, and D is a digraph.

$V(G) : $ the vertex set of G

$E(G) : $ the edge set of G

$v(G) : $ the number of vertices in G or the order of G

$e(G) : $ the number of edges in G or the size of G

$V(D) : $ the vertex set of D

$E(D) : $ the arc set of D

$v(D) : $ the number of vertices in D or the order of D

$e(D) : $ the number of arcs in D

$x \rightarrow y : $ x is adjacent to y, where x, y are vertices in D

$x \not\rightarrow y : $ x is not adjacent to y, where x, y are vertices in D

$G \cong H : $ G is isomorphic to H

$A(G) : $ the adjacency matrix of G

$\overline{G}$: the complement of G

$[A]$: the subgraph of G induced by A, where $A \subseteq V(G)$

$e(A, B)$: the number of edges in G having an end in A and the other in B, where $A, B \subseteq V(G)$

$G - v$: the subgraph of G obtained by removing v and all edges incident with v from G, where $v \in V(G)$

$G - e$: the subgraph of G obtained by removing e from G, where $e \in E(G)$

$G - F$: the subgraph of G obtained by removing all edges in F from G, where $F \subseteq E(G)$

$G - A$: the subgraph of G obtained by removing each vertex in A together with the edges incident with vertices in A from G, where $A \subseteq V(G)$

$G + xy$: the graph obtained by adding a new edge xy to G, where $x, y \in V(G)$ and $xy \notin E(G)$

$N_G(u)$: the set of vertices v such that $uv \in E(G)$

$N(u) = N_G(u)$

$N(S) = \bigcup_{u \in S} N(u)$, where $S \subseteq V(G)$

$d(v) = d_G(v)$: the degree of v in G, where $v \in V(G)$

$id(v)$: the indegree of v in D, where $v \in V(D)$

$od(v)$: the outdegree of v in D, where $v \in V(D)$

$d(u, v)$: the distance between u and v in G, where $u, v \in V(G)$

$d(u, v)$: the distance from u to v in D, where $u, v \in V(D)$

$c(G)$: the number of components in G

$\delta(G)$: the minimum degree of G

$\Delta(G)$: the maximum degree of G

$\chi(G)$: the chromatic number of G

$\alpha(G)$: the independence number of G

$G + H$: the join of G and H

$G \cup H$: the disjoint union of G and H

kG : the disjoint union of k copies of G

$G(D)$: the underlying graph of D

$n_G(H)$: the number of subgraphs in G which are isomorphic to H

C_n : the cycle of order n

K_n : the complete graph of order n

N_n : the null graph or empty graph of order n

P_n : the path of order n

W_n : the wheel of order n, $W_n = C_{n-1} + K_1$

$K(p, q)$: the complete bipartite graph with a bipartition (X, Y) such that $|X| = p$ and $|Y| = q$

Contents

Chapter 1

Fundamental Concepts and Basic Results

Theorem 1.1 *Let G be a multigraph with $V(G) = \{v_1, v_2, \cdots, v_n\}$. Then*

$$\sum_{i=1}^{n} d(v_i) = 2e(G).$$

Corollary 1.2 *The number of odd vertices in any multigraph is even.*

Exercise 1.2

Problem 1. *Let G be the multigraph representing the following diagram. Determine $V(G)$, $E(G)$, $v(G)$ and $e(G)$. Is G a simple graph?*

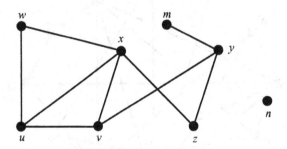

Solution. $V(G) = \{m, n, u, v, w, x, y, z\}$,

$E(G) = \{my, uv, uw, ux, vx, vy, wx, xz, yz\}, \quad v(G) = 8$ and $e(G) = 9$.

Yes, G is a simple graph. □

1

Problem 2. *Draw the graph G modeling the flight connectivity between twelve capital cities with the following vertex set V(G) and edge set E(G).*

$V(G) = \{$*Asuncion, Beijing, Canberra, Dili, Havana, Kuala Lumpur,*
         London, Nairobi, Phnom Penh, Singapore, Wellington,
         Zagreb$\}$.

$E(G) = \{$*Asuncion-Havana, Asuncion-London, Beijing-Canberra,*
         Beijing-Kuala Lumpur, Beijing-London, Beijing-Phnom Penh,
         Beijing-Singapore, Canberra-Dili, Dili-Kuala Lumpur,
         Dili-Singapore, Havana-London, Havana-Nairobi,
         Kuala Lumpur-London, Kuala Lumpur-Phnom Penh,
         Kuala Lumpur-Singapore, Kuala Lumpur-Wellington,
         London-Nairobi, London-Singapore, London-Wellington,
         London-Zagreb, Phnom Penh-Singapore, Singapore-Wellington$\}$.

(Note that you may use 'A' to represent 'Asuncion', 'B' to represent 'Beijing', 'C' to represent 'Canberra', etc.)

Solution.

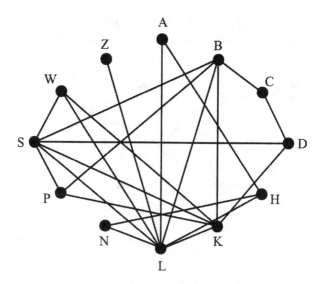

                                                             □

Problem 3. *Define a graph G such that $V(G) = \{2, 3, 4, 5, 11, 12, 13, 14\}$ and two vertices 's' and 't' are adjacent if and only if $gcd\{s, t\} = 1$. Draw a diagram of G and find its size $e(G)$.*

Solution.

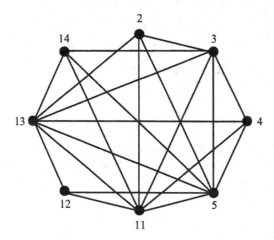

$e(G) = 21.$ ☐

Problem 4. *The diagram below is a map of the road system in a town. Draw a multigraph to model the road system, using a vertex to represent a junction and an edge to represent a road joining two junctions.*

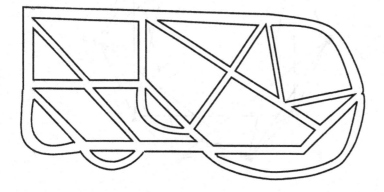

Diagram for Problem 4

Solution.

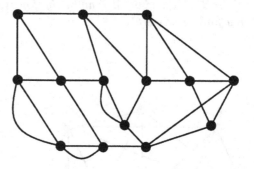

Problem 5. Let G be a graph with $V(G) = \{1, 2, \cdots, 10\}$, such that two numbers 'i' and 'j' in $V(G)$ are adjacent if and only if $|i - j| \leq 3$. Draw the graph G and determine $e(G)$.

Solution.

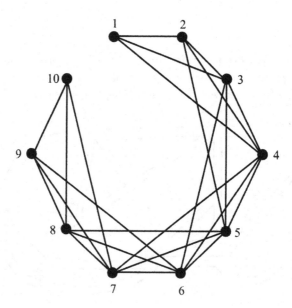

$e(G) = 24.$

Problem 6. *Let G be a graph with $V(G) = \{1, 2, \cdots, 10\}$, such that two numbers 'i' and 'j' in $V(G)$ are adjacent if and only if $i + j$ is a multiple of 4. Draw the graph G and determine $e(G)$.*

Solution.

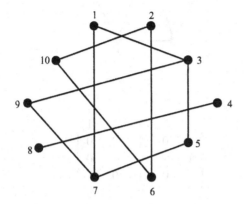

$e(G) = 10.$ □

Problem 7. *Let G be a graph with $V(G) = \{1, 2, \cdots, 10\}$, such that two numbers 'i' and 'j' in $V(G)$ are adjacent if and only if $i \times j$ is a multiple of 10. Draw the graph G and determine $e(G)$.*

Solution.

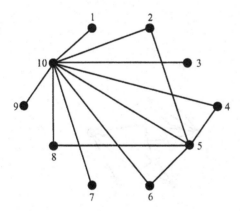

$e(G) = 13.$ □

Problem 8. *Find the adjacency matrix of the following graph G.*

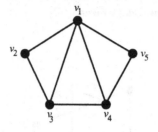

Solution.

$$\begin{pmatrix} 0 & 1 & 1 & 1 & 1 \\ 1 & 0 & 1 & 0 & 0 \\ 1 & 1 & 0 & 1 & 0 \\ 1 & 0 & 1 & 0 & 1 \\ 1 & 0 & 0 & 1 & 0 \end{pmatrix}$$

□

Problem 9. *The adjacency matrix of a multigraph G is shown below:*

$$\begin{pmatrix} 0 & 1 & 0 & 2 & 3 \\ 1 & 0 & 1 & 2 & 2 \\ 0 & 1 & 0 & 1 & 1 \\ 2 & 2 & 1 & 0 & 1 \\ 3 & 2 & 1 & 1 & 0 \end{pmatrix}$$

Draw a diagram of G.

Solution.

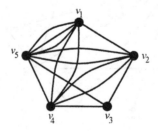

□

Problem 10. *Four teams of three specialist soldiers each (a scout, a signaler and a sniper) are to be sent into enemy territory. However, some of the soldiers cannot work well with some others. The following table shows the soldiers, their specializations and who they cannot work with.*

Soldier	Specialization	Cannot cooperate with
1	Scout	5, 7, 10
2	Scout	–
3	Scout	5, 6, 8, 9, 11
4	Scout	8, 12
5	Signaler	1, 3, 9
6	Signaler	3, 10, 11
7	Signaler	1, 9, 12
8	Signaler	3, 4, 9, 10
9	Sniper	3, 5, 7, 8
10	Sniper	1, 6, 8
11	Sniper	3, 6
12	Sniper	4, 7

(i) *Draw a multigraph to model the situation so that we may see how to form 3-man teams such that each specialization is represented and every member of the team can work with every other. State clearly what the vertices represent and under what condition(s) two vertices are joined by an edge.*

(ii) *Can you form four 3-man teams such that each specialization is represented and all members of the team can work with one another?*

Solution. (i) Vertex *i* represents soldier *i*. Two vertices are joined by an edge if the two corresponding soldiers can cooperate with each other and are not of the same specialization.

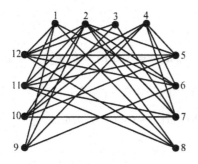

(ii) From the graph, one possible arrangement is

$$\{1,6,9\},\{2,8,12\},\{3,7,10\},\{4,5,11\}.$$

□

Exercise 1.3

Problem 1. *In the following multigraph G, find*

(i) *the size of G,*

(ii) *the degree of each vertex,*

(iii) *the sum* $\sum\limits_{v \in V(G)} d(v),$

(iv) *the number of odd vertices,*

(v) $\Delta(G)$, *and*

(vi) $\delta(G)$.

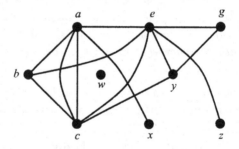

Is your answer for (iii) double your answer for (i)? Is your answer for (iv) an even number?

Solution. (i) $e(G) = 13$.

(ii) $d(a) = 5$, $d(b) = 3$, $d(c) = 5$, $d(e) = 6$, $d(g) = 2$, $d(w) = 0$, $d(x) = 1$, $d(y) = 3$, $d(z) = 1$.

(iii) $\sum\limits_{v \in V(G)} d(v) = 5 + 3 + 5 + 6 + 2 + 0 + 1 + 3 + 1 = 26$.

(iv) There are 6 odd vertices (namely, a, b, c, x, y, z).

(v) $\Delta(G) = 6$.

(vi) $\delta(G) = 0$.

Yes, the answer for (iii) is double that for (i); and the answer for (iv) is an even number. □

Problem 2. *Construct a multigraph of order 6 and size 7 in which every vertex is odd.*

Solution. A required multigraph is shown below.

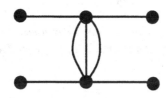

□

Problem 3. *Let G be a multigraph with $V(G) = \{v_1, v_2, \cdots, v_n\}$. Prove that the sum of all the entries in the ith row of the adjacency matrix $A(G)$ is the degree of the vertex v_i for each $i = 1, 2, \cdots, n$.*

Solution. Given i, where $1 \le i \le n$, the sum of the entries in the ith row of $A(G)$ is the sum of the numbers of edges joining v_i to v_j, where $j = 1, 2, \cdots, n$, which is thus the degree of v_i in G. □

Problem 4. *Let G be a graph of order 8 and size 15 in which each vertex is of degree 3 or 5. How many vertices of degree 5 does G have? Construct one such graph G.*

Solution. Let x and y be the number of vertices in G of degree 3 and 5 respectively. Then $x + y = 8$ and $3x + 5y = 2 \times 15 = 30$. Solving the equations yields $(x, y) = (5, 3)$.

An example of G is shown below.

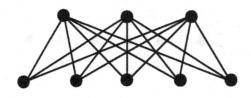

□

Problem 5. *Let H be a graph of order 10 such that $3 \leq d(v) \leq 5$ for each vertex v in H. Not every vertex is even. No two odd vertices are of the same degree. What is the size of H?*

Solution. Let x, y and z be the number of vertices in H of degree $3, 4$ and 5 respectively. Since not every vertex is even, $x + z \geq 2$. As no two odd vertices are of the same degree, $x = z = 1$. Thus, $(x, y, z) = (1, 8, 1)$, and so $e(H) = (3 + 4 \times 8 + 5)/2 = 20$. □

Problem 6. *Let G be a graph of order 14 and size 30 in which every vertex is of degree 4 or 5. How many vertices of degree 5 does G have? Construct one such graph G.*

Solution. Let x and y be the number of vertices in G of degree 4 and 5 respectively. Then $x + y = 14$ and $4x + 5y = 2 \times 30 = 60$. Solving the equations yields $(x, y) = (10, 4)$. Thus, G has 4 vertices of degree 5.

An example of G is shown below.

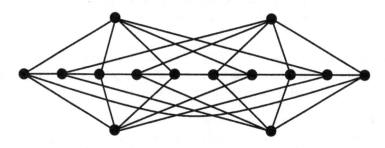

□

Problem 7. *Does there exist a multigraph G of order 8 such that $\delta(G) = 0$ while $\Delta(G) = 7$? What if 'multigraph G' is replaced by 'graph G'?*

Solution. 'Yes' for multigraph G. An example is shown below.

'No' for graph G. Since if there is a vertex v in G with $d(v) = 7$, then v is adjacent to the remaining 7 vertices in G, and so $\delta(G) \geq 1$. $\quad\square$

Problem 8. *Characterize the 1-regular graphs.*

Solution. A graph is 1-regular if and only if it is of even order and is the disjoint union of some K_2's (see below).

$\square$

Problem 9. *Draw all regular graphs of order n, where $2 \leq n \leq 6$.*

Solution. All null graphs N_n and complete graphs K_n, where $2 \leq n \leq 6$, are candidates. The remaining ones are shown below.

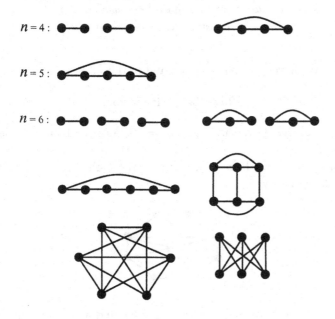

$\square$

Problem 10.

  (i) *Does there exist a graph G of order 5 such that $\delta(G) = 1$ and $\Delta(G) = 4$?*

  (ii) *Does there exist a graph G of order 5 which has two vertices of degree 4 and $\delta(G) = 1$?*

Solution.   (i) Yes. An example is shown below.

(ii) No. Suppose G were such a graph having the vertices u and v of degree 4. As $v(G) = 5$, each of the other vertices must be adjacent to both u and v, and so $\delta(G) \geq 2$.   □

Problem 11. *Let H be a graph of order 8 and size 13 with $\delta(H) = 2$ and $\Delta(H) = 4$. Denote by n_i the number of vertices in H of degree i, where $i = 2, 3, 4$. Assume that $n_3 \geq 1$. Find all possible answers for (n_2, n_3, n_4). For each of your answers, construct a corresponding graph.*

Solution.   We have $n_2 + n_3 + n_4 = 8$ and by Theorem 1.1,

$$2n_2 + 3n_3 + 4n_4 = 26.$$

It follows from the above that $n_3 + 2n_4 = 10$. As $n_3 \geq 1$, by Corollary 1.2, $n_3 = 2, 4$ or 6.

    When $n_3 = 2$, we have $(n_2, n_3, n_4) = (2, 2, 4)$, and a corresponding graph is shown below:

When $n_3 = 4$, we have $(n_2, n_3, n_4) = (1, 4, 3)$, and a corresponding graph is shown below:

When $n_3 = 6$, we have $n_4 = 2$, and so $n_2 = 0$, which is not possible as $\delta(H) = 2$.   □

Problem 12. *Suppose G is a k-regular graph of order n and size m, where $k \geq 0$, $m \geq 0$ and $n \geq 1$. Find a relation linking k, n and m. Justify your answer.*

Solution.   By Theorem 1.1, $kn = \sum\limits_{x \in V(G)} d(x) = 2m$.   □

Problem 13. *Does there exist a 3-regular graph with eight vertices? Does there exist a 3-regular graph with nine vertices?*

Solution.   Yes, a 3-regular graph of order 8 is shown below.

No, there does not exist any 3-regular graph of order 9 by Corollary 1.2 (or the result of Problem 12).   □

Problem 14. *Construct a cubic (i.e., 3-regular) graph of order 12. What is its size? Does there exist a cubic graph of order 11? Why?*

Solution.   A cubic graph of order 12 is shown below.

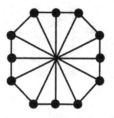

Its size is $(3 \times 12)/2 = 18$ (see Problem 12).

By Corollary 1.2 (or the result of Problem 12), there does not exist any cubic graph of order 11.   □

Problem 15. *Let H be a k-regular graph of order n. If $e(H) = 10$, find all possible values for k and n; and for each case, construct one such graph H.*

Solution. By the result of Problem 12, $kn = 20$. As $k \leq n - 1$,

$$(k, n) = (1, 20), (2, 10) \text{ or } (4, 5).$$

An example of H for each case is shown below:

□

Problem 16. (+) *Let G be a 3-regular graph with $e(G) = 2v(G) - 3$. Determine the values of $v(G)$ and $e(G)$. Construct all such graphs G.*

Solution. Let $n = v(G)$ and $m = e(G)$. By the assumption and Theorem 1.1, we have: $3n = 2m = 2(2n - 3)$, which implies that $(n, m) = (6, 9)$. There are only two such G as shown below.

□

Problem 17. *Find all integers n such that $100 \leq e(K_n) \leq 200$.*

Solution. As $e(K_n) = n(n-1)/2$, we have $200 \leq n(n-1) \leq 400$. It follows that $15 \leq n \leq 20$. □

Problem 18. (+) *Let G be a multigraph of order 13 in which each vertex is of degree 7 or 8. Show that G contains **at least eight** vertices of degree 7 or **at least seven vertices** of degree 8.*

Solution. Suppose that the conclusion is **false**. Then G contains **at most seven** vertices of degree 7 **and at most six** vertices of degree 8. Since $v(G) = 13$, G contains **exactly seven** vertices of degree 7. This is, however, impossible by Corollary 1.2. □

Problem 19. (+) *Let G be a graph of order n in which there exist **no three** vertices u, v and w such that uv, vw and wu are all edges in G. Show that $n \geq \delta(G) + \Delta(G)$.*

Solution. Let x be a vertex in G such that $d(x) = \Delta(G)$. Pick a vertex y in $N(x)$. By assumption,

$$n \geq 1 + \Delta(G) + (d(y) - 1) = \Delta(G) + d(y) \geq \Delta(G) + \delta(G).$$

□

Problem 20. (+) *There were n (≥ 2) persons at a party and, as usually happens, some shook hands with others. No one shook hands with the same person more than once. Show that there are at least two persons in the party who had the same number of handshakes.*

Solution. Model the situation as a graph G of order n, where the vertices are the persons, and two vertices are adjacent if and only if the two corresponding persons shook hands. By assumption, G is a simple graph. The problem is equivalent to showing that there exist two vertices u, v in G such that $d(u) = d(v)$.

It is clear that $0 \leq d(x) \leq n - 1$ for each vertex x in G. If the above statement is false, then there exist two vertices y and z in G such that $d(y) = 0$ and $d(z) = n - 1$, which however is impossible. □

Problem 21. *The preceding problem says that in any graph of order $n \geq 2$, there exist two vertices having the same degree. Is the result still valid for multigraphs?*

Solution. No! A multigraph in which no two vertices have the same degree is shown below.

□

Problem 22. (+) *Mr. and Mrs. Samy attended an exclusive party where in addition to themselves, there were only another 3 couples. As usually happens, some shook hands with others. No one shook hands with the same person more than once and no one shook hands with his/her spouse. After all the handshakes had been done, Mr. Samy asked each person, including his wife, how many hands he/she had shaken. To everyone's amusement, each one gave a different answer. How many hands did Mrs. Samy shake?*

Solution. Model the situation by a graph G with 8 vertices for 8 persons, and defining 'adjacency' for 'handshaking'. By assumption, $0 \leq d(v) \leq 6$ for each v in G, and each of '0, 1, 2, 3, 4, 5, 6' is the *degree* of some vertex.

Let v_1 be such that $d(v_1) = 6$ and $N(v_1) = \{v_2, v_3, \cdots, v_7\}$, say. Then $d(v_8) = 0$, and v_1 and v_8 are spouses (see below).

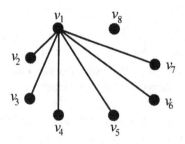

Let v_2 be such that $d(v_2) = 5$ and $N(v_2) = \{v_1, v_3, v_4, v_5, v_6\}$, say. Then $d(v_7) = 1$ and v_2 and v_7 are spouses (see below).

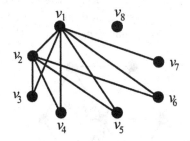

Let v_3 be such that $d(v_3) = 4$ and $N(v_3) = \{v_1, v_2, v_4, v_5\}$, say. Then $d(v_6) = 2$ and v_3 and v_6 are spouses (see below).

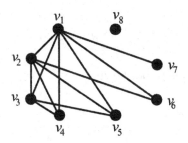

It follows that $d(v_4) = d(v_5) = 3$, and v_4 and v_5 are spouses.

As Mr Samy received different answers, either v_4 or v_5 represents Mrs Samy. Thus Mrs Samy shook hands with three others. □

Problem 23. (+) *In the preceding problem, there were four couples alto-gether in a party. Solve the general problem where 'four couples' is replaced by 'n(≥ 2) couples'.*

Solution. Using a similar argument as shown in the solution of Problem 22, it can be shown that the answer is '$n - 1$' for this general problem. □

Problem 24. (∗) *There are $n \geq 2$ distinct points in the plane such that the distance between any 2 points is at least one. Prove that there are at most $3n$ pairs of these points at distance exactly one.*

Solution. Let $p_1, p_2, \cdots, p_n$ be the n given points in the plane. Form a graph G with $V(G) = \{p_1, p_2, \cdots, p_n\}$ in which two vertices are adjacent if their distance in the plane is '1'. What is the largest possible value that each $d(p_i)$ can have? By the assumption that the distance between any 2 points is at least one, it follows that $d(p_i) \leq 6$ (see the figure below).

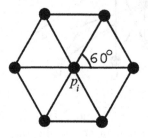

Thus, by Theorem 1.1,

$$2e(G) = \sum_{i=1}^{n} d(p_i) \leq 6n,$$

and so $e(G) \leq 3n$, as was to be shown.   □

Exercise 1.4

Problem 1. *Consider the following graph H.*

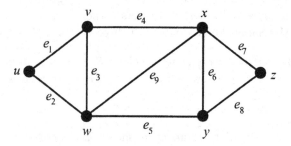

(a) *Which of the following sequences represents a $u - z$ walk in H?*
  (i) $ue_2we_5xe_7z$
  (ii) $ue_1ve_5ye_8z$
  (iii) $ue_1ve_3we_3ve_4xe_7z$

(b) *Find a $u - z$ trail in H that is not a path.*

(c) *Find all $u - z$ paths in H which pass through e_9.*

Solution. (a) Only the sequence (iii) represents a $u - z$ walk in H.

    (b) The sequence "$uvxywxz$" is a $u - z$ trail that is not a path.

    (c) All such paths are: $uvwxz, uvwxyz, uwxz, uwxyz, uvxwyz$.   □

Problem 2. *Consider the following multigraph G:*

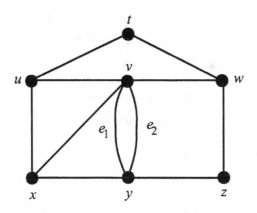

(a) Find $d(t, v)$, $d(t, y)$, $d(x, w)$ and $d(u, z)$.

(b) For $k = 2, 3, 4, 5, 6, 7$, find a cycle of length k in G.

(c) Find a circuit of length 6 in G that is not a cycle.

(d) Find a circuit of length 8 in G that does not contain t.

(e) Find a circuit of length 9 in G that contains t and v.

Solution.

(a) $d(t, v) = 2, d(t, y) = 3, d(x, w) = 2$ and $d(u, z) = 3$.

(b) ve_1ye_2v is a cycle of length 2,

    $uvxu$ is a cycle of length 3,

    $utwvu$ is a cycle of length 4,

    $xvwzyx$ is a cycle of length 5,

    $uvwzyxu$ is a cycle of length 6 and

    $uvxyzwtu$ is a cycle of length 7.

(c) $tuve_1ye_2vwt$ is a circuit of length 6 that is not a cycle.

(d) $uvwzye_1ve_2yxu$ is a circuit of length 8 that does not contain 't'.

(e) $utwzye_1ve_2yxvu$ is a circuit of length 9 that contains both 't' and 'v'. □

Problem 3. *Is the following graph H disconnected? If it is so, find its number of components.*

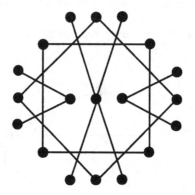

Solution.   Yes, H is disconnected, and it has 5 components.   □

Problem 4. *Let G be a graph with $V(G) = \{1, 2, \cdots, n\}$, where $n \geq 5$, such that two numbers i and j in $V(G)$ are adjacent if and only if $|i - j| = 5$. How many components does G have?*

Solution.   By definition, the graph G is depicted as follows:

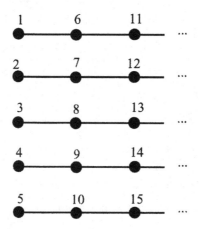

Thus, G has 5 components.   □

Problem 5. (+) *Show that any $u - v$ walk in a graph contains a $u - v$ path.*

Solution. Let P be a $u - v$ walk. We may assume that $u \neq v$. If no vertex in P is repeated, then P is a path, and we are through. Assume that a vertex x is repeated in P as shown below (it is possible that $x = u$ or $x = v$):

$$P : \underbrace{u \cdots}_{(a)} \underbrace{x \cdots}_{(b)} \underbrace{x \cdots v}_{(c)}.$$

Then P can be cut short by deleting the section (b) internally resulting in a shorter $u - v$ walk P' as shown below:

$$P' : \underbrace{u \cdots}_{(a)} \underbrace{x \cdots v}_{(c)}.$$

This procedure is repeatedly applied until no vertex in the resulting $u - v$ walk is repeated, and in this case, the resulting $u - v$ walk is a desired path. $\square$

Problem 6. (+) *Show that any circuit in a graph contains a cycle.*

Solution. Let Q be a circuit of length at least 2. If no vertex in Q is repeated, then Q is a cycle, and we are through. Assume that a vertex x is repeated in Q as shown below:

$$P : \underbrace{u \cdots}_{(a)} \underbrace{x \cdots}_{(b)} \underbrace{x \cdots u}_{(c)}.$$

Then Q can be cut short by deleting the section (b) internally resulting in a shorter circuit Q' as shown below:

$$Q' : \underbrace{u \cdots}_{(a)} \underbrace{x \cdots u}_{(c)}.$$

This procedure is repeatedly applied until no vertex in the resulting circuit is repeated, and in this case, the resulting circuit is a desired cycle. $\square$

Problem 7. (+) *Show that any graph G with $\delta(G) \geq k$ contains a path of length k.*

Solution. Let $P = v_0 v_1 \cdots v_r$ be a longest path (of length r) in G. By assumption, $d(v_0) \geq k$, and so v_0 has at least k neighbors. Note that all these neighbors must be contained in P; for if there is a neighbor (say, w) of v_0 not in P, then we would have a path of the form: $w v_0 v_1 \cdots v_r$, which is of length $r + 1$, contradicting the fact that P is a longest path. Thus, $N(v_0) \subseteq \{v_1, \cdots, v_r\}$, and so $r \geq |N(v_0)| = d(v_0) \geq k$, as required. $\square$

Problem 8. (+) *Let G be a graph of order $n \geq 2$ such that $\delta(G) \geq \frac{1}{2}(n-1)$. Show that $d(u, v) \leq 2$ for any two vertices u, v in G.*

Solution. Let u, v be any two distinct vertices u, v in G. If u and v are adjacent, then $d(u, v) = 1$. Assume that u and v are not adjacent. Consider $N(u)$ and $N(v)$. We claim that $N(u) \cap N(v) \neq \emptyset$.

Suppose that $N(u) \cap N(v) = \emptyset$. Then, as $\{u, v\} \cup N(u) \cup N(v) \subseteq V(G)$ and $\delta(G) \geq (n - 1)/2$, we have:

$$n \geq 2 + |N(u)| + |N(v)| = 2 + d(u) + d(v) \geq 2 + 2\delta(G) \geq n + 1,$$

which is impossible.

Thus, $N(u) \cap N(v) \neq \emptyset$, as claimed. Let $w \in N(u) \cap N(v)$. Then uwv is a $u - v$ path of length 2, and so $d(u, v) = 2$.

We thus conclude that $d(u, v) \leq 2$, for any two vertices u, v in G. $\square$

Problem 9. (+) *Let G be a graph of order n and size m such that $m > \binom{n-1}{2}$. Show that G is connected.*

Solution. Suppose on the contrary that G is disconnected. Let G_1 be a component of order k ($1 \leq k \leq n - 1$) in G, and let G_2 be the remaining part of G, which is of order $n - k$. Then

$$\binom{n-1}{2} < m = e(G_1) + e(G_2) \leq \binom{k}{2} + \binom{n-k}{2},$$

which implies that

$$(n - 1)(n - 2) < k(k - 1) + (n - k)(n - k - 1)$$

or $(k - (n - 1))(k - 1) > 0$. As $k \geq 1$, it follows that $k - (n - 1) > 0$, i.e. $k > n - 1$, a contradiction.

We thus conclude that G is connected if the condition holds. $\square$

Problem 10. *For $n \geq 2$, construct a disconnected graph of order n and size $\binom{n-1}{2}$.*

Solution. The graph with 2 components, namely, K_1 and K_{n-1}, is the candidate. □

Problem 11. *Let G be a disconnected graph of order 5. What is the largest possible value for $e(G)$? If G is a disconnected graph of order $n \geq 2$, what is the largest possible value for $e(G)$? Construct one such extremal graph of order n.*

Solution. If G is a disconnected graph of order $n \geq 2$, the largest possible value for $e(G)$ is $\binom{n-1}{2}$.

To justify this, we note that the disconnected graph $K_1 \cup K_{n-1}$ has order n and its size equal to $\binom{n-1}{2}$, and by Problem 9 above, there is no disconnected graph of order n having its size greater than $\binom{n-1}{2}$. □

Problem 12. (+) *Let G be a graph of order $n \geq 2$ and u, v be two non-adjacent vertices in G such that $d(u) + d(v) \geq n + r - 2$. Show that u and v have at least r common neighbours.*

Solution. Our aim is to show that $|N(u) \cap N(v)| \geq r$. By the Principle of Inclusion and Exclusion, we have

$$|N(u) \cap N(v)| = |N(u)| + |N(v)| - |N(u) \cup N(v)|$$
$$= d(u) + d(v) - |N(u) \cup N(v)|.$$

As u and v are non-adjacent, $N(u) \cup N(v) \subseteq V(G) \backslash \{u, v\}$. Thus,

$$|N(u) \cap N(v)| = d(u) + d(v) - |N(u) \cup N(v)|$$
$$\geq d(u) + d(v) - |V(G) \backslash \{u, v\}|$$
$$\geq n + r - 2 - (n - 2) \quad \text{(by assumption)}$$
$$= r,$$

as was shown. □

Problem 13. (+) *Let G be a connected graph that is not complete. Show that there exist three vertices x, y, z in G such that x and y, y and z are adjacent, but x and z are not adjacent in G.*

Solution. As G is not complete, let u and v be two non-adjacent vertices in G. As G is connected, let $uab\cdots v$ be a **shortest** $u-v$ path in G (note that it is possible that $b=v$). Now let $x=u, y=a$ and $z=b$. As $uab\cdots v$ is a **shortest** $u-v$ path, it follows that x and z are not adjacent in G. $\square$

Problem 14. (+) *Let G be a graph of order n and size m such that $\Delta(G) = n-2$ and $d(u,v) \le 2$ for any two vertices u,v in G. Show that $m \ge 2n-4$.*

Solution. Let x be a vertex in G such that $d(x) = \Delta(G) = n-2$ with $N(x) = \{y_1, y_2, \cdots, y_{n-2}\}$ as shown below:

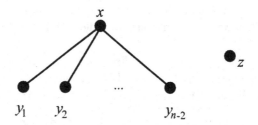

Clearly, the $(n-2)$ xy_i's are edges in G. As G is of order n, let z be the remaining vertex in G. Note that z and x are not adjacent.

Since $d(z,x) = 2$ by assumption, z must be adjacent to some y_i's. Without loss of generality, we may assume that z is adjacent to $y_1, y_2, \cdots, y_k$, where $1 \le k \le n-2$ and k is the largest index such that z is adjacent to y_k.

Now, for each j (if any) with $k+1 \le j \le n-2$, as $d(z, y_j) = 2$ (z and y_j are not adjacent now), there must be a new edge joining y_j with some y_i in $\{y_1, y_2, \cdots, y_k\}$.

Summing up, the number of edges in G is at least

$$(n-2) + k + ((n-2) - k) = 2n - 4.$$

That is, $m \ge 2(n-2)$, as required. $\square$

Problem 15. *Let G be a graph such that $N(x) \cup N(y) = V(G)$ for every pair of vertices x, y in G. What can be said of G?*

Solution. The graph G must be a complete graph. We justify it as follows. Suppose that G is not complete. Then there exist two non-adjacent vertices u and v in G. In this case, u is not contained in $N(u) \cup N(v)$; which, however, contradicts the assumption that $N(u) \cup N(v) = V(G)$. $\square$

Problem 16. (+) *Let H be a graph of order $n \geq 2$. Suppose that H contains two distinct vertices u, v such that (i) $N(u) \cup N(v) = V(H)$ and (ii) $N(u) \cap N(v)$ is non-empty.*

What is the least possible value of $e(H)$?

Solution. The least possible size of H is 'n' (note that the given conditions imply that $n \geq 3$). The justification is as follows.

Firstly, the following graph of order n satisfying the given conditions contains exactly 'n' edges:

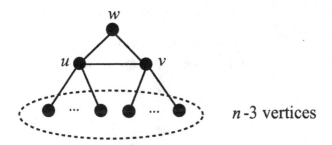

Now we show that every graph H satisfying the given conditions must have at least n edges.

The vertices u and v must be adjacent; for if not, then u is not in $N(u) \cup N(v)$, and so $N(u) \cup N(v) \neq V(H)$, violating the condition (i).

By (ii), there exists a vertex, say w, in $N(u) \cap N(v)$, and so w is adjacent to both u and v.

By (i) again, each of the $(n-3)$ vertices other than u, v and w must be adjacent to either u or v.

Summing up, H contains at least $1 + 2 + (n-3)$ edges; that is, $e(H) \geq n$, as required.   □

Problem 17. *Suppose G is a disconnected graph which contains exactly two odd vertices u and v. Must u and v be in the same component of G? Why?*

Solution. Yes, the two odd vertices u and v must be in the same component of G. Otherwise, let H be the component of G containing u but not v; then H is a graph containing exactly one odd vertex, contradicting Corollary 1.2.   □

Problem 18. (∗) *Show that any two longest paths in a connected graph have a vertex in common.*

Solution. Let P and Q be two longest paths (of length k each, say) in a connected graph G, and suppose on the contrary that they have no vertex in common. As G is connected, there exist a vertex u in P and a vertex v in Q which are joined by a path R, say. Without loss of generality, we may assume (see the figure below) that (i) this $u - v$ path R contains no vertex in P or Q other than u and v, (ii) the length of the subpath (a) in P is greater than or equal to that of (b) and (iii) the length of the subpath (d) in Q is greater than or equal to that of (c).

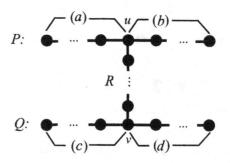

With this, however, we observe that the path consisting of the subpath (a) in P, the $u - v$ path R and the subpath (d) in Q is of length greater than k, a contradiction.   □

Problem 19. (+) *Show that a graph G is connected if and only if for any partition of $V(G)$ into two non-empty sets A and B, there is an edge in G joining a vertex in A and a vertex in B.*

Solution. [Necessity] Suppose on the contrary that there is a partition (A, B) of $V(G)$ for which there is no edge in G joining a vertex in A and a vertex in B. It is then clear that no vertex in A can be joined to any vertex in B by a path. Thus G is disconnected, a contradiction.

     [Sufficiency] Suppose on the contrary that G is disconnected. Let H be a component of G and R the remaining part of G. Then $(V(H), V(R))$ forms a partition of $V(G)$ for which there is no edge in G joining a vertex in $V(H)$ and a vertex in $V(R)$ (note that both $V(H)$ and $V(R)$ are non-empty), a contradiction.   □

Problem 20. (∗) *Suppose G is a connected graph with k edges. Prove that it is possible to label the edges $1, 2, \cdots, k$ in such a way that at each vertex which belongs to two or more edges (i.e. which is of degree at least two), the greatest common divisor of the integers labeling those edges is 1 (32nd IMO, 1991/4). (Recall that the greatest common divisor of the positive integers $x_1, x_2, \cdots, x_n$ is the maximum positive integer that divides each of $x_1, x_2, \cdots, x_n$.)*

Proof. Starting at an arbitrary vertex, say v_0, in G, we walk along distinct edges in G to produce a maximal trail (no edge is repeated), say of length s, and label the edges along the trail $1, 2, \cdots, s$ (see the example below).

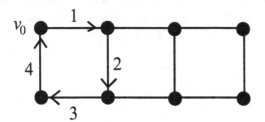

If there are edges not yet labeled, as G is connected, one of them is incident with a vertex, say v_r, which has been visited. Starting at v_r, we walk along distinct unlabelled edges in G to produce another maximal trail, say of length p, and label the edges along the trail $s + 1, s + 2, \cdots, s + p$ (see the diagram below).

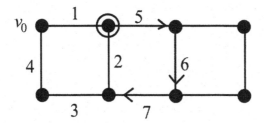

We repeat the above procedure until all edges in G are labeled (see the diagram below).

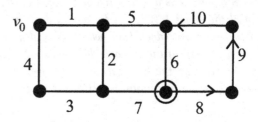

We now show that for each vertex v with $d(v) \geq 2$ in G, the *gcd* of the labels of the edges incident with v is 1. If $v = v_0$, the situation is clear as the first edge incident with it is labeled 1. Assume that $v \neq v_0$. Let e be the edge with which we first visit v via a trail. As $d(v) \geq 2$, along the same trail, we leave v with a new edge, say f. By the above procedure, the labels of e and f are consecutive numbers, say t and $t+1$, and so the corresponding *gcd* (that is, $gcd\{t, t+1, \cdots\}$) is 1.   □

Chapter 2

Isomorphisms, Subgraphs and the Complement of a Graph

Result (1). If $G \cong H$, then $v(G) = v(H)$ and $e(G) = e(H)$.

Result (2). If $G \cong H$, then G and H have the same degree sequence, in non-increasing order.

Result (3). Let G and H be graphs such that $G \cong H$. Then for any graph R, $n_G(R) = n_H(R)$.

Result (4). For any graph G of order n, $e(G) + e(\overline{G}) = e(K_n) = \binom{n}{2}$.

Result (5). Let G be a graph. If G is disconnected, then $\overline{G}$ is connected.

Result (6). Let G be a self-complementary graph. Then
(i) G is connected and
(ii) $v(G) = 4k$ or $v(G) = 4k + 1$ for some integer k.

Exercise 2.2

Problem 1. *Draw all non-isomorphic graphs of order n with* $1 \leq n \leq 4$.

Solution.

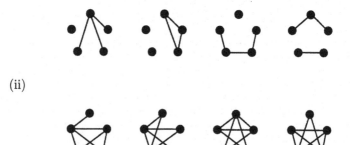

Problem 2.

(i) *Draw all non-isomorphic graphs of order 5 and size 3.*

(ii) *Draw all non-isomorphic graphs of order 5 and size 7.*

Solution. (i)

(ii)

Problem 3. *Determine if the following two graphs are isomorphic.*

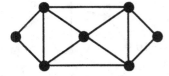   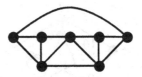

Solution.
No, since the degree sequences $(4, 4, 4, 4, 4, 2, 2)$ and $(4, 4, 4, 3, 3, 3, 3)$ are not the same.

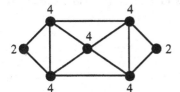   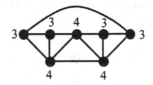

□

Problem 4. *Determine if the following two graphs are isomorphic.*

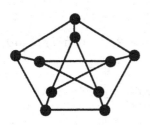   

Solution.   Yes. As shown below, let

$$V(G) = \{1, 2, \cdots, 10\} \text{ and } V(H) = \{1', 2', \cdots, 10'\}.$$

It can be checked that the mapping $f : V(G) \longrightarrow V(H)$, such that $f(i) = i'$ for all $i = 1, 2, \cdots, 10$, is an isomorphism from G to H.

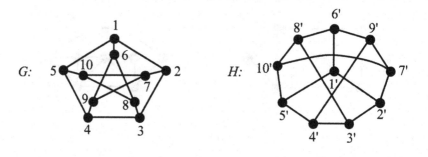

Problem 5. *The following two graphs G and H are isomorphic. List all the isomorphisms from G to H.*

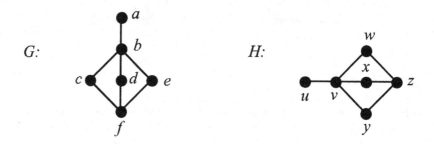

Solution.   There are six isomorphisms $g_i, i = 1, 2, \cdots, 6$, from G to H:

$$g_i(a) = u, \ g_i(b) = v, \ g_i(f) = z, \quad i = 1, 2, \cdots, 6,$$

$$g_1(c) = w, g_1(d) = x, g_1(e) = y,$$

$$g_2(c) = w, g_2(d) = y, g_2(e) = x,$$

$$g_3(c) = x, g_3(d) = w, g_3(e) = y,$$

$$g_4(c) = x, g_4(d) = y, g_4(e) = w,$$

$$g_5(c) = y, g_5(d) = w, g_5(e) = x,$$

$$g_6(c) = y, g_6(d) = x, g_6(e) = w.$$

Problem 6. (∗) *Prove, by definition of an isomorphism, that the relation '≅' is reflexive, symmetric and transitive among the family of graphs; that is, the properties listed in Question 2.1.3.*

Solution. (i) $G \cong G$ since the identity mapping is one-to-one and onto, and clearly preserves adjacency.

(ii) Suppose $G \cong H$. Then there exists a one-to-one and onto mapping f from $V(G)$ to $V(H)$ which preserves adjacency. Clearly, f^{-1} is a one-to-one and onto mapping from $V(H)$ to $V(G)$ which preserves adjacency. Thus, $H \cong G$.

(iii) Suppose $G \cong H$ and $H \cong J$. Then there exist one-to-one and onto mappings, f from $V(G)$ to $V(H)$ and g from $V(H)$ to $V(J)$, which preserve adjacency. Now, the composite mapping $g \circ f$ is a one-to-one and onto mapping from $V(G)$ to $V(J)$ which preserves adjacency. Thus, $G \cong J$. □

Problem 7. *Let f be an isomorphism from a graph G to a graph H and w a vertex in G. Show that the degree of w in G is equal to the degree of $f(w)$ in H.*

Solution. Let the set of vertices adjacent to w in G be $\{v_1, v_2, \cdots, v_k\}$. Then the set of vertices adjacent to $f(w)$ in H is $\{f(v_1), f(v_2), \cdots, f(v_k)\}$. Hence

$$d(w) = |\{v_1, v_2, \cdots, v_k\}| = k = |\{f(v_1), f(v_2), \cdots, f(v_k)\}| = d(f(w)).$$

□

Problem 8. *A given graph G of order 5 contains at least two vertices of degree 4.*

(i) *Assume that not all vertices in G are even. Find all possible degree sequences of G, in non-increasing order; and for each case, construct all such G which are not isomorphic.*

(ii) *Assume that all vertices in G are even. Find all possible degree sequences of G, in non-increasing order; and for each case, construct all such G which are not isomorphic.*

Solution. (i) Since there are at least two vertices of degree 4, the other three vertices must have degree at least 2. Since not all vertices are even and we know that there must be an even number of odd vertices, there are exactly two vertices of degree 3.

Thus, the possible degree sequences are $(4, 4, 4, 3, 3)$ and $(4, 4, 3, 3, 2)$. Their graphs are shown below.

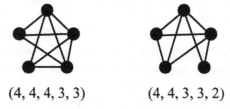

(4, 4, 4, 3, 3)   (4, 4, 3, 3, 2)

(ii) Since there are at least two vertices of degree 4, the other three vertices must have degree at least 2. Since all vertices are even, if there is among these three vertices, one of degree 4, then all the vertices are of degree 4. Otherwise, all the other three vertices are of degree 2.

Thus, the possible degree sequences are $(4, 4, 4, 4, 4)$ and $(4, 4, 2, 2, 2)$. Their graphs are shown below.

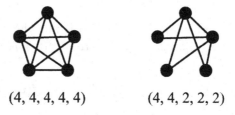

(4, 4, 4, 4, 4)   (4, 4, 2, 2, 2)

$\square$

Problem 9. *Let H be a graph of order 5 which contains more odd vertices than even. Find all possible degree sequences of H in non-increasing order; and for each case, construct all such H which are not isomorphic.*

Solution. The possible degree sequences are

$$(1, 1, 1, 1, 0), (2, 1, 1, 1, 1), (3, 1, 1, 1, 0), (3, 2, 1, 1, 1), (3, 3, 2, 1, 1),$$

$$(3, 3, 3, 2, 1), (3, 3, 3, 3, 0), (3, 3, 3, 3, 2), (4, 1, 1, 1, 1) \text{ and } (4, 3, 3, 3, 3).$$

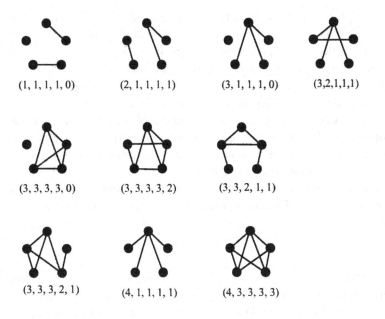

(1, 1, 1, 1, 0) (2, 1, 1, 1, 1) (3, 1, 1, 1, 0) (3,2,1,1,1)

(3, 3, 3, 3, 0) (3, 3, 3, 3, 2) (3, 3, 2, 1, 1)

(3, 3, 3, 2, 1) (4, 1, 1, 1, 1) (4, 3, 3, 3, 3)

□

Problem 10. *Construct two non-isomorphic* 3-*regular graphs of order* 10.

Solution. The following two 3-regular graphs of order 10 are not isomorphic:

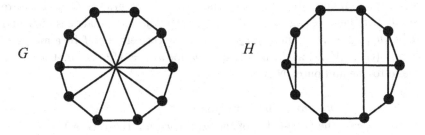

Note that H contains a C_3 while G does not. □

Problem 11. *Let G and H be two isomorphic graphs. Show that*

(i) *if G is connected, then H is connected;*

(ii) *if G is disconnected, then H is disconnected, and they have the same number of components.*

Solution. (i) Let u and v be any two vertices in H. Let f be an isomorphism from H to G. Since G is connected, there is in G an $f(u) - f(v)$ path, say, $f(u)f(u_1)f(u_2)\cdots f(u_k)f(v)$ where $u_i \in V(H)$ for all $i = 1, 2, \cdots, k$. Then $uu_1u_2\cdots u_kv$ is a path in H. Thus, H is connected.

(ii) By (i), if H is connected, then G is connected. Thus by the contrapositive statement, if G is disconnected, then H is disconnected.

Suppose that G and H have different numbers of components, say G has g components and H has less than g components. Let $G_1, G_2, \cdots, G_g$ be the components of G. Select vertices v_i, $i = 1, 2, \cdots, g$, such that $v_i \in V(G_i)$. Let f be an isomorphism from G to H.

Suppose, for some i, j with $1 \le i < j \le g$, $f(v_i)$ and $f(v_j)$ are in the same component of H. Then there exists in H a $f(v_i) - f(v_j)$ path

$$f(u_0)f(u_1)\cdots f(u_t)$$

for some vertices $u_0, u_1, u_2, \cdots, u_t \in V(G)$. where $u_0 = v_i$, $u_t = v_j$ and $t \ge 1$. However, $u_0u_1\cdots u_t$ is a $v_i - v_j$ path in G, a contradiction. Thus, any two of the g vertices $f(v_1), f(v_2), \cdots, f(v_g)$ are not in the same component of H, implying that H contains at least g components, a contradiction. $\square$

Problem 12. *Prove that if the adjacency matrices of two graphs G and H are equal, then the graphs G and H are isomorphic.*

Solution. Suppose the adjacency matrices of two graphs G and H are equal. Let the ordering of the vertices in the adjacency matrices of G and H be $(u_1, u_2, \cdots, u_n)$ and $(v_1, v_2, \cdots, v_n)$ respectively. Let f be a mapping from $V(G)$ to $V(H)$ such that $f(u_i) = v_i$ for all $i = 1, 2, \cdots, n$. Clearly, f is one-to-one and onto. Now,

$$u_i \text{ and } u_j \text{ in } G \text{ are adjacent}$$
$$\iff \text{the } (i, j)\text{-entry of the adjacency matrix of } G \text{ is } 1$$
$$\iff \text{the } (i, j)\text{-entry of the adjacency matrix of } H \text{ is } 1$$
$$\iff v_i \text{ and } v_j \text{ in } H \text{ are adjacent.}$$

Thus, f preserves adjacency and so the graphs G and H are isomorphic. $\square$

Problem 13. *Using adjacency matrices, determine which, if any, of the following three graphs are isomorphic.*

 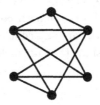

Solution. Let us label the vertices of the three graphs (which we shall name A, B and C) as in the figure below.

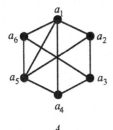 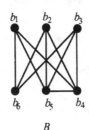 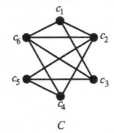

| A | B | C |

Adjacency matrices for the three graphs are shown below.

$$
\begin{array}{c}
a_1\ a_2\ a_3\ a_4\ a_5\ a_6 \\
A_1 = \begin{pmatrix}
0 & 1 & 0 & 1 & 1 & 1 \\
1 & 0 & 1 & 0 & 1 & 0 \\
0 & 1 & 0 & 1 & 0 & 1 \\
1 & 0 & 1 & 0 & 1 & 0 \\
1 & 1 & 0 & 1 & 0 & 1 \\
1 & 0 & 1 & 0 & 1 & 0
\end{pmatrix}
\end{array}
\qquad
\begin{array}{c}
b_1\ b_2\ b_3\ b_4\ b_5\ b_6 \\
B_1 = \begin{pmatrix}
0 & 0 & 0 & 1 & 1 & 1 \\
0 & 0 & 0 & 1 & 1 & 1 \\
0 & 0 & 0 & 1 & 1 & 1 \\
1 & 1 & 1 & 0 & 1 & 0 \\
1 & 1 & 1 & 1 & 0 & 0 \\
1 & 1 & 1 & 0 & 0 & 0
\end{pmatrix}
\end{array}
\qquad
\begin{array}{c}
c_1\ c_2\ c_3\ c_4\ c_5\ c_6 \\
C_1 = \begin{pmatrix}
0 & 1 & 1 & 0 & 0 & 1 \\
1 & 0 & 0 & 1 & 1 & 1 \\
1 & 0 & 0 & 0 & 1 & 1 \\
0 & 1 & 0 & 0 & 1 & 1 \\
0 & 1 & 1 & 1 & 0 & 0 \\
1 & 1 & 1 & 1 & 0 & 0
\end{pmatrix}
\end{array}
$$

We now rearrange the rows (and correspondingly the columns) of A_1 such that the vertices in the new matrix A_2 are now in the order $a_2, a_4, a_6, a_5, a_1, a_3$.

$$
\begin{array}{c}
a_2\ a_4\ a_6\ a_5\ a_1\ a_3 \\
A_2 = \begin{pmatrix}
0 & 0 & 0 & 1 & 1 & 1 \\
0 & 0 & 0 & 1 & 1 & 1 \\
0 & 0 & 0 & 1 & 1 & 1 \\
1 & 1 & 1 & 0 & 1 & 0 \\
1 & 1 & 1 & 1 & 0 & 0 \\
1 & 1 & 1 & 0 & 0 & 0
\end{pmatrix}
\end{array}
$$

Since $A_2 = B_1$, we have that $A \cong B$.

We observe that three vertices in B, namely b_1, b_2, b_3, are such that no two of them are adjacent to each other. A systematic check of the vertices of C will show that no such three vertices exist. Thus, $C \not\cong B$.   □

Exercise 2.3

Problem 1. *Let G be the graph given below:*

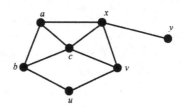

(i) *Draw the subgraphs $[\{b, v, y\}]$, $[\{a, b, c, v, x\}]$ and $[\{a, b, u, v, x\}]$ of G.*
(ii) *Draw the subgraphs $G - \{ab, cv, xy\}$, $G - c$ and $G - \{b, v\}$ of G.*
(iii) *Find $E([\{a, b, c, x\}])$.*
(iv) *Draw the subgraph $G - E([\{a, b, c, x\}])$ of G.*
(v) *Draw a spanning subgraph of G that is connected and that contains a unique C_3 as a subgraph.*
(vi) *Draw a spanning subgraph of G that is connected and that contains no cycle as a subgraph.*

Solution.   (i) The subgraphs $[\{b, v, y\}]$, $[\{a, b, c, v, x\}]$ and $[\{a, b, u, v, x\}]$ of G are the following graphs respectively.

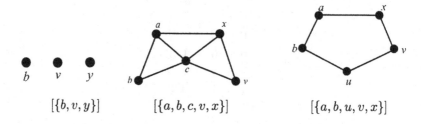

(ii) The subgraphs $G - \{ab, cv, xy\}$, $G - c$ and $G - \{b, v\}$ of G are the following graphs respectively.

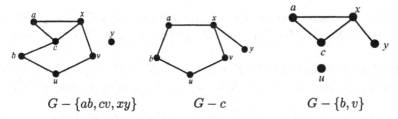

$$G - \{ab, cv, xy\} \qquad G - c \qquad G - \{b, v\}$$

(iii) $E([\{a, b, c, x\}]) = \{ab, ac, ax, bc, cx\}$.

(iv) The subgraph $G - E([\{a, b, c, x\}])$ is

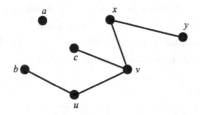

(v) The following subgraph of G is spanning and connected, and it contains a unique C_3:

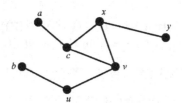

(vi) The following subgraph of G is spanning and connected, and it contains no cycles:

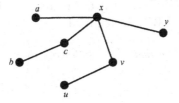

□

Problem 2. *Let H be a subgraph of a graph G. Show that H is a spanning subgraph of G if and only if $H = G - F$, where $F \subseteq E(G)$.*

Solution. ($\Rightarrow$) Assume that H is a spanning subgraph of G. Then $V(H) = V(G)$ and $E(H) \subseteq E(G)$. Let $F = E(G) \backslash E(H)$. Clearly, $H = G - F$, where $F \subseteq E(G)$.

($\Leftarrow$) Suppose that $H = G - F$, where $F \subseteq E(G)$. Clearly, $V(H) = V(G)$ and $E(H) \subseteq E(G)$. Thus, by definition, H is a spanning subgraph of G. $\square$

Problem 3. *Let G be a graph and $X \subseteq V(G)$. Show that $G - X = [V(G) \backslash X]$.*

Solution. We shall show that $G - X$ is the subgraph of G induced by $V(G) \backslash X$, where $X \subseteq V(G)$. We first note that $V(G - X) = V(G) \backslash X$. Next, we have to show that if there is an edge in G joining two vertices in $V(G) \backslash X$, then this edge must be in $G - X$. Indeed, if $e = uv$ is such an edge in G joining vertices u and v in $V(G) \backslash X$, then as u and v are not in X, the edge 'e' is still an edge in $G - X$.

We thus conclude that $G - X = [V(G) \backslash X]$. $\square$

Problem 4. *Let G be a graph and W a subgraph of G. Show that W is an induced subgraph of G if and only if $W = G - (V(G) \backslash V(W))$.*

Solution. ($\Leftarrow$) Assume that $W = G - (V(G) \backslash V(W))$.

Then by letting $V(G) \backslash V(W) = X$ in the result of Problem 3, we have

$$W = G - (V(G) \backslash V(W)) = G - X = [V(G) \backslash X] = [V(W)];$$

that is, W is an induced subgraph of G.

($\Rightarrow$) Suppose that W is an induced subgraph of G. Then

$$W = [V(W)] = [V(G) \backslash (V(G) \backslash V(W))],$$

which is equal to $G - (V(G) \backslash V(W))$ by the result in Problem 3. Thus we have $W = G - (V(G) \backslash V(W))$, as required. $\square$

Problem 5. *Determine which of the following four graphs are isomorphic and which are not so.*

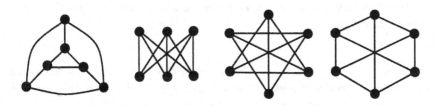

Solution. Denote the four graphs by G_1, G_2, G_3 and G_4 (from left to right) respectively. Then $G_1 \cong G_3$ and $G_2 \cong G_4$, and their isomorphisms are shown below.

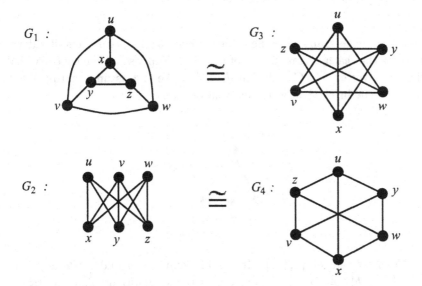

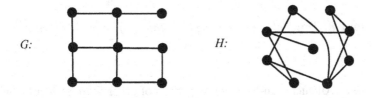

Note that G_1 is not isomorphic to G_2 (thus G_3 is not isomorphic to G_4) as, for instance, G_1 contains a triangle while G_2 does not. □

Problem 6. *Let G and H be the two graphs given below:*

Do they have the same degree sequence in non-increasing order? Are they isomorphic?

Solution. Yes, the graphs G and H have the same degree sequence in non-increasing order, namely, $(4, 3, 3, 3, 2, 2, 2, 2, 1)$. However, they are not isomorphic as, for instance, H contains a triangle while G does not.   □

Problem 7. (+) *Let G be a graph of order five satisfying the following condition: for any three vertices x, y, z in G, $[\{x, y, z\}]$ is not isomorphic to either N_3 or K_3.*
    What is the graph G? Justify your answer.

Solution. The graph G is the cycle of order 5. To justify this, it suffices to show that each vertex in G is of degree 2. Suppose on the contrary that there is a vertex v in G such that $d(v) \neq 2$. We first assume that $d(v) \geq 3$. Let a, b, c be in $N(v)$ (see the diagram below).

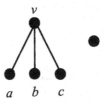

Consider the set $\{a, b, c\}$. If two of them, say a and b, are adjacent, then $[\{v, a, b\}] \cong K_3$, which contradicts the assumption. Thus, no two in $\{a, b, c\}$ are adjacent, which, however, implies that $[\{a, b, c\}] \cong N_3$, again a contradiction.
    The case that $d(v) = 1$ likewise leads to a contradiction (we leave it to the reader). Thus $d(v) = 2$ for each vertex v in G, as claimed.
Note. The reader should argue why $G \cong C_5$ in this case.   □

Problem 8. *Draw all non-isomorphic graphs of order 5 which contain a C_5.*

Solution. All such non-isomorphic graphs of order 5 are shown below:

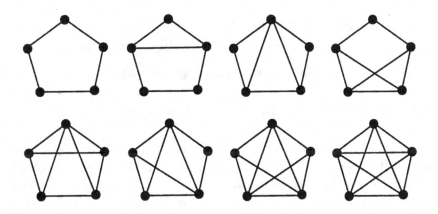

□

Problem 9. *Let H be a spanning subgraph of a graph G. Which of the following statements is/are true? Why?*

(i) *If G is connected, then H is connected.*

(ii) *If H is connected, then G is connected.*

Solution. (i) False. N_n is a spanning subgraph of K_n. While K_n is always connected, N_n is not connected if $n \geq 2$.

(ii) True. Assume that H is connected. We now prove that G is connected by showing that every two vertices in G are joined by a path. Thus, let u and v be any two vertices in G. Since H is a spanning subgraph of G, u and v are also vertices in H. As H is connected, there is a $u - v$ path, say P, in H. Since H is a subgraph of G, we have $V(H) \subseteq V(G)$ and $E(H) \subseteq E(G)$, and so P is also a $u - v$ path in G. This shows that u and v are joined by a path in G. □

Problem 10. *Let G be a disconnected graph with k components. Choose a vertex from each component. What is the subgraph induced by these k vertices?*

Solution. The subgraph of G induced by these k vertices is isomorphic to N_k as no two of these vertices are adjacent in G. □

Problem 11. *For a graph G, denote by c(G) the number of components in G. Thus, for the graph G below, c(G) = 4.*

G: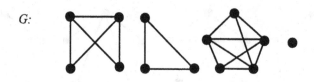

Let H be a spanning subgraph of a graph G. Show that $c(H) \geq c(G)$.

Solution. Suppose that $c(G) = k$ and let $G_1, \cdots, G_k$ be the components of G with vertex sets $V_1, \cdots, V_k$ respectively. Write $H[V_i]$ for the subgraph of H induced by V_i. Clearly, for each $i = 1, 2, \cdots, k$, $H[V_i]$ contains at least one component of H, and these components are different. Thus, $c(H) \geq k = c(G)$.   □

Problem 12. (+) *Let C be a cycle and S a subset of $V(C)$. Show that $c(C - S) \leq |S|$.*

Solution. Draw the cycle C as shown below, where the vertices of S are denoted by big black dots and each of the components of $C - S$ is enclosed.

Define a mapping f from the set of components of $C - S$ to the set S as follows: for each component 'A' of $C - S$, let $f(A)$ be the first vertex in S next to it clockwise.

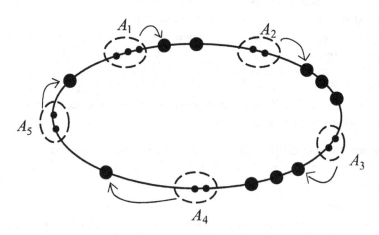

The mapping f is clearly one-to-one. Thus, $c(C - S) \leq |S|$.   □

Problem 13. *Let $G = K_n$. Find*

(i) $n_G(C_3)$, where $n \geq 3$;

(ii) $n_G(C_4)$, where $n \geq 4$; and

(iii) $n_G(C_k)$, where $n \geq k \geq 5$.

Solution. (i) Every three vertices in K_n induce a C_3. Thus the answer is $\binom{n}{3}$.

(ii) Every four vertices in K_n induce a K_4, which, in turn, contains exactly three C_4 as shown below:

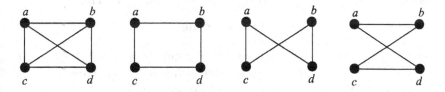

Thus the answer is $3\binom{n}{4}$.

(iii) For $5 \leq k \leq n$, every k vertices in K_n induce a K_k, and each K_k contains exactly $(k-1)!/2$ distinct C_k. Thus the answer is $(k-1)!\binom{n}{k}/2$. □

Problem 14. *Let G be the Petersen graph. Find $n_G(C_i)$, where $i = 3, 4, 5$. What is the largest cycle in G?*

Solution. $n(C_3) = n(C_4) = 0$ and $n(C_5) = 12$.

The Petersen graph (of order 10) contains no spanning cycle. Its largest cycle is of order 9 as shown below:

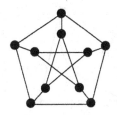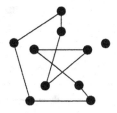

□

Problem 15. *Let G be a graph of order 5 which contains at least two vertices of degree 4 and a C_5. Find all possible degree sequences of G, in* *non-increasing order; and for each case, construct all such G.*

Solution. By assumption, G contains the following C_5 as a spanning subgraph:

As G contains at least two vertices of degree 4, $e(G) \geq 8$. Thus

$$8 \leq e(G) \leq 10.$$

When $e(G) = 10$, $G = K_5$ and its degree sequence is $(4, 4, 4, 4, 4)$. When $e(G) = 9$, $G = K_5 - e$, where e is an edge in K_5, and its degree sequence is $(4, 4, 4, 3, 3)$. When $e(G) = 8$, G is the following graph:

and its degree sequence is $(4, 4, 3, 3, 2)$.   □

Problem 16. *Let G be a connected graph. An edge e in G is called a* **bridge** *if $G - e$ is disconnected.*

  (i) *Find all bridges in the following graph:*

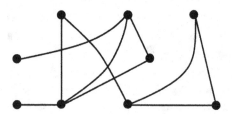

  (ii) *How many components does $G - e$ have if e is a bridge in G?*

 (iii) *Show that an edge e in G is a bridge if and only if e is not contained in any cycle in G.*

Solution. (i) There are four bridges as indicated below:

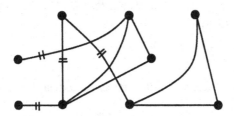

(ii) $c(G - e) = 2$ if e is a bridge in G.

(iii) Let e be an edge in a connected graph G.

($\Leftarrow$) Assume that e is not contained in any cycle. We shall show that e is a bridge. Suppose that $e(= uv)$ is not a bridge. Then, by definition, $G - e$ is connected. Thus, u and v, being vertices in $G - e$, are joined by a path P in $G - e$. Now, in G, the path P together with the edge e forms a cycle containing e, a contradiction.

($\Rightarrow$) Assume now that e is a bridge in G. We shall show that e is not contained in any cycle. Since $e(= uv)$ is a bridge, $G - e$ is disconnected. Thus, there is no path joining u and v in $G - e$, which implies that e is contained in no cycle in G. $\qquad\square$

Problem 17. (+) *Let G be a connected graph in which every vertex is even. Show that G contains no bridges.*

Solution. Suppose on the contrary that G contains a bridge $e(= uv)$. Let H be the component of $G - e$ containing the vertex u. Then, by assumption, u is the only odd vertex in the graph H, contradicting Corollary 1.2. $\qquad\square$

Problem 18. *Let G be a connected graph. A vertex w in G is called a* **cut-vertex** *if $G - w$ is disconnected.*

(i) *Find all cut-vertices in the following graph:*

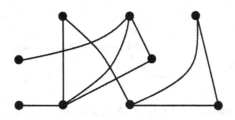

(ii) *How many components does $G - w$ have if w is a cut-vertex of G?*

(iii) *(+) Assume that $v(G) \geq 3$. Show that if G contains a bridge, then G contains a cut-vertex.*

(iv) *Is the converse of (iii) true?*

Solution.   (i) There are four cut-vertices as indicated below:

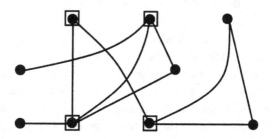

(ii) The value of $c(G - w)$ can be any positive number if w is a cut vertex of G as shown below:

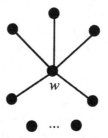

(iii) Let $e(= uv)$ be a bridge in G. As G is connected and $v(G) \geq 3$, either u or v (say v) is adjacent to a vertex w other than u (see the diagram below). It is clear that u and w are not joined by any path in $G - v$. Thus $G - v$ is disconnected, and by definition, v is a cut-vertex in G.

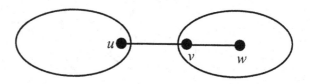

(iv) The converse is not true. An example is given below:

☐

Problem 19. (∗) *Let G be a cubic (i.e., 3-regular) graph. Suppose G contains a cut-vertex. Must G contain a bridge? Why?*

Solution. Yes, G must contain a bridge in this case. The justification is given below. Let w be a cut-vertex in G. As G is cubic (i.e. 3-regular), w has exactly 3 neighbors, say x, y and z (see the diagrams below). Observe that $G - w$ is disconnected and one of its components contains exactly one of the x, y and z, say z. In this case, wz is a bridge in G.

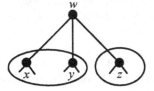 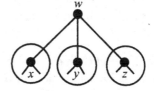

☐

Problem 20. *Let G be a connected graph of order 8 and size 12 which contains no bridges. Suppose that $\Delta(G) = 4$ and G has exactly two vertices of degree 4.*

(i) *Find the number of end-vertices in G.*

(ii) *Find the number of vertices of degree 3 in G.*

(iii) *Construct three such graphs which are non-isomorphic.*

Solution. (i) As G contains no bridges, G contains no end-vertices.

(ii) Let x and y denote, respectively, the number of vertices of degree 2 and 3. Then $x + y = 6$ and, by Theorem 1.1, $2x + 3y + 4 \times 2 = 24$. Solving these equations yields $(x, y) = (2, 4)$.

(iii) Three examples of such graphs are shown below:

Problem 21. (+) *Show that a graph G contains a cycle of length at least $\delta(G) + 1$ if $\delta(G) \geq 2$.*

Solution. Let $\delta(G) = k(\geq 2)$ and let $P = v_0v_1 \cdots v_r$ be a longest path (of length r) in G. As $d(v_0) \geq k$, v_0 has at least k neighbors. Note that all these neighbors must be in P; for if there is a neighbor (say, w) of v not in P, then we would have a path of the form: $wv_0v_1 \cdots v_r$, which is of length $r + 1$, contradicting the fact that P is a longest path.

Let s be the largest index, $2 \leq s \leq r$, such that v_s is adjacent to v_0. Clearly, $s \geq k$ and $v_0v_1 \cdots v_sv_0$ is a cycle of length $s + 1(\geq k + 1)$, as required. □

Problem 22. *Let G be a graph of order 9. Assume that $\Delta(G) = 6$ and that G contains at least 4 vertices of degree at least 4. Show that G contains a C_3.*

Solution. Let w be a vertex in G with $d(w) = \Delta(G) = 6$. Since G has at least four vertices of degree at least 4, one of the vertices in $N(w)$ must be of degree at least 4. Call this vertex u (see the diagram below). As $d(u) \geq 4$, u must be adjacent to another vertex in $N(w)$, say v. Clearly, $wuvw$ forms a triangle.

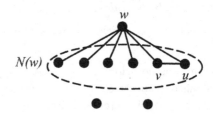

□

Problem 23. *Let G be a graph of order n with degree sequence $(d_1, d_2, \cdots, d_n)$. Construct a graph from G having the degree sequence $(d_1 + 1, d_2 + 1, \cdots, d_n + 1, n)$.*

Solution. Given a graph G of order n with degree sequence $(d_1, d_2, \cdots, d_n)$, let H be the graph obtained by adding a new vertex w to G and joining w to every vertex in G (see the diagram below). It can be checked that the degree sequence of H is $(d_1 + 1, d_2 + 1, \cdots, d_n + 1, n)$.

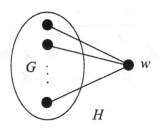

□

Problem 24. (+) *Let G be a connected graph of order n. Show that the vertices in G can always be named as $x_1, x_2, \cdots, x_n$ such that the induced subgraph $[\{x_1, x_2, \cdots, x_i\}]$ is connected for each $i = 1, 2, \cdots, n$.*

Solution. Start by picking a vertex at random and naming it x_1. Pick any unnamed neighbor of x_1, and name it x_2. In general, having named vertices with the names $x_1, x_2, \cdots, x_k$, check through all neighbors of x_k. If there is such a vertex unnamed, pick one and name it x_{k+1}. Otherwise, find the largest index j such that x_j has an unnamed neighbor. Pick such an unnamed neighbor, and name it x_{k+1}.

An example of the above procedure is shown below. For convenience, we denote x_j by 'j'.

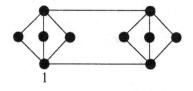

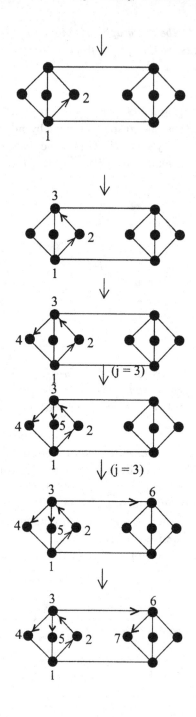

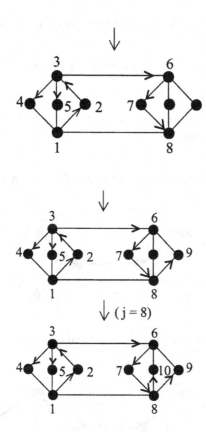

We shall now prove that the subgraph $[\{x_1, x_2, \cdots, x_i\}]$ is connected for $i = 1, 2, \cdots, n$ by induction. The statement is obvious for $i = 1, 2$. Assume that it is true for $i = k$. Consider the case that $i = k + 1$. We know from the procedure that the vertex x_{k+1} is a neighbor of either x_k or x_j, where $1 \leq j \leq k$. As $[\{x_1, x_2, \cdots, x_k\}]$ is connected by the induction hypothesis, it follows that $[\{x_1, x_2, \cdots, x_{k+1}\}]$ is also connected.

This completes the proof. □

Problem 25. *Let G be a connected graph of order 8 which contains two C_4's having no vertex in common.*

(i) *What is the least possible value of $e(G)$?*

(ii) *Assume that G contains no cut-vertices. What is the least possible value of $e(G)$?*

(iii) *Assume that G contains no odd vertices. What is the least possible value of e(G)?*

(iv) *Assume that G contains no even vertices. What is the least possible value of e(G)?*

For each of the above cases, construct a corresponding G which has its e(G) attaining your least possible value.

Solution. (i) The least possible value of $e(G)$ is 9. An example is shown below:

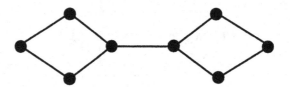

(ii) If G contains no cut-vertices, the least possible value of $e(G)$ is 10. An example is shown below:

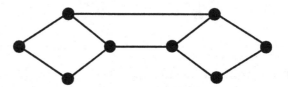

(iii) If G contains no odd vertices, the least possible value of $e(G)$ is 11. An example is shown below:

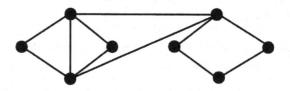

(iv) If G contains no even vertices, the least possible value of $e(G)$ is 12. Two examples are shown below:

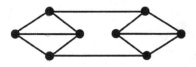 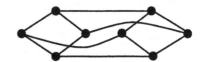

□

Problem 26. *Let G be a graph with* $V(G) = \{x_1, x_2, x_3, x_4\}$ *such that*
$G-x_1 \cong$ 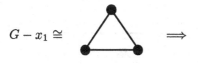 , $G-x_2 \cong$ • •——• , $G-x_3 \cong$ •——•——• *and* $G-x_4 \cong$
•——•——• .
Determine G and justify your answer.

Solution. Notation:

  $H \subseteq G$: H is a subgraph of G.

  u •┄┄┄• v : u and v are not adjacent.

$G - x_1 \cong$ 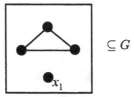 $\Longrightarrow$ $\subseteq G$

$G - x_2 \cong$ • •——•   (*by symmetry*) $\Longrightarrow$ 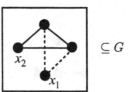 $\subseteq G$

$G - x_3 \cong$ •——•——•   (*by symmetry*) $\Longrightarrow$ 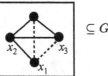 $\subseteq G$

It follows that G is the graph:

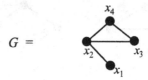

$$G =$$

$\square$

Problem 27. (+) *Let G be a graph with $V(G) = \{y_1, y_2, \cdots, y_5\}$ such that*

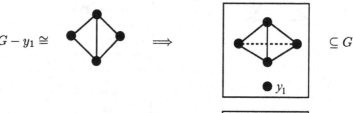

$G - y_1 \cong$, $G - y_2 \cong$, $G - y_3 \cong$, $G - y_4 \cong$ *and*

$G - y_5 \cong$.

Determine G and justify your answer.

Solution.

$G - y_1 \cong$ $\implies$ $\subseteq G$

$G - y_2 \cong$ (by symmetry) $\implies$ $\subseteq G$

$G - y_3 \cong$ (by symmetry) $\implies$ $\subseteq G$

$$G - y_4 \cong \quad\Longrightarrow\quad = G$$

Problem 28. (+) *Let G be a graph with $V(G) = \{u_1, u_2, \cdots, u_n\}$, where $n \geq 3$. Let $m = e(G)$, $m_i = e(G - u_i)$, $i = 1, 2, \cdots, n$. Show that*

(i) *the degree of u_i in G is equal to $m - m_i$, $i = 1, 2, \cdots, n$;*

(ii) *$m = (m_1 + m_2 + \cdots + m_n)/(n - 2)$.*

Solution. (i) As shown in the following diagram, it is clear that, for each $i = 1, 2, \cdots, n$,

$$e(G) = e(G - u_i) + d(u_i).$$

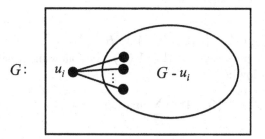

Thus, $d(u_i) = e(G) - e(G - u_i) = m - m_i$, for each $i = 1, 2, \cdots, n$.

(ii) Observe that

$$2m = \sum_{i=1}^{n} d(u_i) \qquad \text{(Theorem 1.1)}$$

$$= \sum_{i=1}^{n} (m - m_i) \qquad \text{(Part (i))}$$

$$= nm - \sum_{i=1}^{n} m_i.$$

Thus, $\sum_{i=1}^{n} m_i = nm - 2m = m(n - 2)$, as required. $\qquad\square$

Problem 29. (+) *Let G be a connected multigraph of order at least two and A be a subset of $V(G)$. Denote by $e(A, V(G)\backslash A)$ the number of edges having one end in A and the other in $V(G)\backslash A$.*

(i) *Let H be the multigraph shown below and $A = \{u, v, z\}$. Find $e(A, V(H)\backslash A)$.*

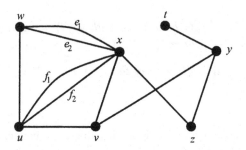

(ii) *Show that $e(A, V(G)\backslash A)$ is even if and only if A contains an even number of odd vertices in G.*

Solution. (i) For $A = \{u, v, z\}$, $e(A, V(H)\backslash A) = 7$.

(ii) We first note that for $A \subseteq V(G)$, the following equality holds (see the diagram below):

$$\sum_{x \in A} d(x) = e(A, V(G)\backslash A) + \sum_{x \in A} d_{[A]}(x) \tag{1}$$

where $d_{[A]}(x)$ denotes the degree of x in $[A]$.

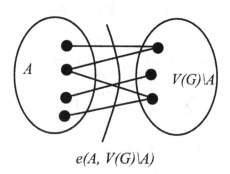

$$e(A, V(G)\backslash A)$$

Let A_e and A_o be, respectively, the set of even vertices and the set of odd

vertices in A. Then

$$\sum_{x \in A} d(x) = \sum_{x \in A_e} d(x) + \sum_{x \in A_o} d(x).$$

By Theorem 1.1, the sum $\sum_{x \in A} d_{[A]}(x)$ is always even. It thus follows from (1) that

$$e(A, V(G) \backslash A) \text{ is even} \iff \sum_{x \in A} d(x) \text{ is even}$$

$$\iff \sum_{x \in A_e} d(x) + \sum_{x \in A_o} d(x) \text{ is even}$$

$$\iff \sum_{x \in A_o} d(x) \text{ is even}$$

$$\iff |A_o| \text{ is even.}$$

The proof is thus complete. $\square$

Exercise 2.4

Problem 1. *Consider Problem 2 in Exercise 2.2. Is there any relation between the family of graphs found in (i) and the family of graphs in (ii)?*

Solution. Yes, there is a one-to-one correspondence between these two families. Indeed, each graph in the family (i) has its complement in the family (ii), and vice versa. $\square$

Problem 2.

(i) *Draw all non-isomorphic graphs of order 6 and size 3.*

(ii) *Find the number of non-isomorphic graphs of order 6 and size 12.*

Solution. (i) There are five such graphs as shown below:

(ii) There are also five such graphs as shown below:

Note that each graph in (i) has its complement in (ii).   □

Problem 3. *Let G and H be two graphs. Show that $G \cong H$ if and only if $\overline{G} \cong \overline{H}$.*

Solution. ($\Rightarrow$) Assume that $G \cong H$. We shall show that $\overline{G} \cong \overline{H}$.

Let $f : G \cong H$ be an isomorphism. We claim that f is also an isomorphism from $\overline{G}$ onto $\overline{H}$. As $V(\overline{G}) = V(G)$ and $V(\overline{H}) = V(H)$, $f : V(\overline{G}) \longrightarrow V(\overline{H})$ is also a one-to-one and onto mapping. Let u, v be in $V(\overline{G})$. Observe that

$$u \text{ and } v \text{ are adjacent in } \overline{G} \Leftrightarrow u \text{ and } v \text{ are not adjacent in } G$$
$$\Leftrightarrow f(u) \text{ and } f(v) \text{ are not adjacent in } H$$
$$\Leftrightarrow f(u) \text{ and } f(v) \text{ are adjacent in } \overline{H}.$$

This shows that f is an isomorphism from $\overline{G}$ onto $\overline{H}$, as claimed.

($\Leftarrow$) Assume that $\overline{G} \cong \overline{H}$. Then, by the above result, we have $\overline{\overline{G}} \cong \overline{\overline{H}}$; that is, $G \cong H$ (see Question 2.4.1(3)).   □

Problem 4. *Let G be a disconnected graph. Show that the distance between any two vertices in $\overline{G}$ is at most two. (See the proof of Result (5).)*

Solution. Assume that G is a disconnected graph. Let $u, v \in V(\overline{G})(= V(G))$. We shall show that $d(u, v) \leq 2$ in $\overline{G}$.

(1) If u, v are in different components in G, then u and v are joined by an edge in $\overline{G}$ and so $d(u, v) = 1$ in $\overline{G}$.

(2) If u and v are in the same component of G, let w be any vertex in another component of G, then uwv is a $u - v$ path in $\overline{G}$ (see Figure 2.22), and so $d(u, v) = 2$ in $\overline{G}$.

This completes the proof.   □

Problem 5. *Let G be a k-regular graph of order n. Is $\overline{G}$ also regular? If the answer is 'yes', what is the degree of each vertex in $\overline{G}$?*

Solution. Let v be a vertex in G. Then $d(v) = k$, and so the degree of v in $\overline{G}$ is $n - 1 - k$. Thus, if G is k-regular, then $\overline{G}$ is $(n - 1 - k)$-regular. $\square$

Problem 6. *Let G be a graph of order n with degree sequence $(d_1, d_2, \cdots, d_n)$ in non-increasing order. Find the degree sequence of $\overline{G}$ in non-increasing order.*

Solution. Let $d_1 \geq d_2 \geq \cdots \geq d_n$ be the non-increasing degree sequence of G. Then $n - 1 - d_n \geq n - 1 - d_{n-1} \geq \cdots \geq n - 1 - d_1$ is the non-increasing degree sequence of $\overline{G}$. (See Problem 5 above.) $\square$

Problem 7. *Draw all non-isomorphic 4-regular graphs of order 7.*

Solution. All non-isomorphic 2-regular graphs of order 7 can be easily constructed as shown below (two such graphs):

By taking their complements, we obtain all non-isomorphic 4-regular graphs of order 7 as shown below:

$\square$

Problem 8. *How many non-isomorphic graphs are there with degree sequence $(5, 5, 4, 4, 4, 4)$? Construct one such graph.*

Solution. We shall apply the result in Problem 6 above and thus first consider the degree sequence $(1, 1, 1, 1, 0, 0)$. It is easy to see that the following graph is the only graph with this degree sequence:

Thus, by taking its complement, the only graph with degree sequence $(5, 5, 4, 4, 4, 4)$ is shown below:

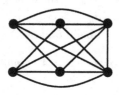

□

Problem 9. *How many non-isomorphic graphs are there with degree sequence* $(5, 5, 5, 4, 4, 3)$*? Construct one such graph.*

Solution. We shall apply the result in Problem 6 above again and thus first consider the degree sequence $(2, 1, 1, 0, 0, 0)$. It is easy to see that the following graph is the only graph with this degree sequence:

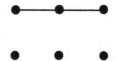

Thus, by taking its complement, the only graph with degree sequence $(5, 5, 5, 4, 4, 3)$ is shown below:

□

Problem 10. (+) *What is the value of each diagonal entry in the matrix* $A(G)A(\overline{G})$*?*

Solution. Let $A(G) = (a_{i,j})$, $A(\overline{G}) = (\overline{a_{i,j}})$ and $A(G)A(\overline{G}) = (b_{i,j})$. We shall determine the value of $b_{i,i}$, $i = 1, 2, \cdots, n$.

Note that for $i \neq j$, $a_{i,j} = 0$ if and only if $\overline{a_{i,j}} = 1$. Thus

$$b_{i,i} = (a_{i,1}, a_{i,2}, \cdots, a_{i,n}) \begin{pmatrix} \overline{a_{1,i}} \\ \overline{a_{2,i}} \\ \cdots \\ \overline{a_{n,i}} \end{pmatrix}$$

$$= (a_{i,1}, a_{i,2}, \cdots, a_{i,n}) \begin{pmatrix} \overline{a_{i,1}} \\ \overline{a_{i,2}} \\ \cdots \\ \overline{a_{i,n}} \end{pmatrix}$$

$$= a_{i,1}\overline{a_{i,1}} + a_{i,2}\overline{a_{i,2}} + \cdots + a_{i,n}\overline{a_{i,n}}$$

$$= 0.$$

$\square$

Problem 11. *For each of the following graphs,*

(i) *construct its complement and*

(ii) *determine if it is self-complementary.*

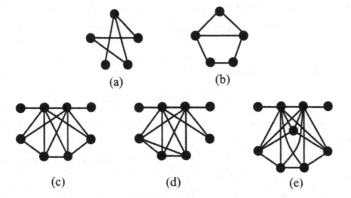

(a)   (b)

(c)   (d)   (e)

Solution. (i) The complements of these five graphs are shown below:

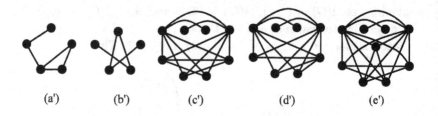

(a')   (b')   (c')   (d')   (e')

(ii) It can be verified that only the graphs of (c) and (e) are self-complementary.

The graph of (a) contains no C_3, but its complement does.

The graph of (b) contains 6 edges, but its complement has only 4 edges.

The graph of (d) contains a vertex of degree 2, but its complement does not.   □

Problem 12. (+) *Show that every self-complementary graph is connected.*

Solution. Let G be a self-complementary graph (i.e. $G \cong \overline{G}$). We shall show that G is connected. Suppose on the contrary that G is disconnected. Then, by Result (5), $\overline{G}$ is connected, which however contradicts the fact that $G \cong \overline{G}$. (See Problem 11(i) in Exercise 2.2.) Thus G is connected. □

Problem 13. (+) *Let G be a self-complementary graph of order $n \geq 2$. Show that*

(i) $e(G) = \frac{1}{4}n(n-1)$ *and*

(ii) $n = 4k$ *or* $n = 4k+1$ *for some positive integer k.*

Solution. (i) By Result (4),

$$e(G) + e(\overline{G}) = n(n-1)/2.$$

As G is self-complementary, $e(G) = e(\overline{G})$. It follows from these two equalities that $e(G) = n(n-1)/4$.

(ii) By (i), we have $e(G) = n(n-1)/4$. Since $e(G)$ is a whole number, the product $n(n-1)$ is divisible by 4.

If n is odd, then $n-1$ must be divisible by 4, and so $n-1 = 4k$ or $n = 4k+1$ for some positive integer k.

If n is even, then $n-1$ is odd, and so n must be divisible by 4. It follows that $n = 4k$ for some positive integer k.   □

Problem 14. *Determine the values of n, where n ≥ 3, for which the cycle C_n is self-complementary.*

Solution. We have: $e(C_n) = n$ and $e(\overline{C_n}) = n(n-1)/2 - n$.
 If C_n is self-complementary, then $e(C_n) = e(\overline{C_n})$, that is,

$$n = n(n-1)/2 - n.$$

Upon simplification, we have $n(n-5) = 0$, and so $n = 5$.
 It is known that C_5 is self-complementary. Thus, we conclude that the cycle C_n is self-complementary if and only if $n = 5$. □

Problem 15. *Let G be a self-complementary graph of order n. Show that if G is regular, then $n = 4k + 1$ for some positive integer k.*

Solution. Let G be an r-regular and self-complementary graph of order n. By Problem 5 above, $\overline{G}$ is $(n - 1 - r)$-regular.
 As $G \cong \overline{G}$, $r = n - 1 - r$, and so $n = 2r + 1$, which is odd. By Problem 13(ii) above, $n = 4k + 1$ for some positive integer k. □

Problem 16. *Construct a regular self-complementary graph of order 9.*

Solution. A regular self-complementary graph of order 9 is shown below:

□

Problem 17. (+) *(i) Let G be a self-complementary graph of order 9. Show that G contains at least one vertex of degree 4.*
 (ii) Generalize the result in (i).

Solution. (i) We shall prove the result by two different methods.
 Method 1. Let $d_1 \geq d_2 \geq \cdots \geq d_9$ be the non-increasing degree sequence of G. Then $8 - d_9 \geq 8 - d_8 \geq \cdots \geq 8 - d_1$ is the non-increasing degree

sequence of $\overline{G}$ (see Problem 6 above). As $G \cong \overline{G}$, these two sequences are identical. In particular, their fifth terms must be equal; that is, $d_5 = 8 - d_5$. It follows that $d_5 = 4$, as required.

Method 2. We first observe that if x is a vertex of degree r in G, then x is of degree $8 - r$ in $\overline{G}$. As $G \cong \overline{G}$, there must exist a vertex, say x', of degree $8 - r$ in G too (see the diagram below). Now, suppose on the contrary that G contains no vertex of degree 4. Then $r \neq 8 - r$, and it follows from the above observation that the vertices in G can be grouped in pairs $\{x, x'\}$. This implies that the order of G is even, a contradiction.

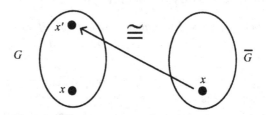

(ii) The general result is as follows:

If G is a self-complementary graph of order $4k + 1$, then G contains at least one vertex of degree $2k$.

Note. The reader is encouraged to generalize the above arguments to prove this general result.   □

Problem 18. *Let G be a graph and x be a vertex in G.*

(i) *Is it true that $\overline{G - x} = \overline{G} - x$?*

(ii) *If x is a cut-vertex of G, is $\overline{G} - x$ connected?*

Justify your answers.

Solution.   (i) Yes, for any vertex x in G, it is true that $\overline{G - x} = \overline{G} - x$.

First of all, it is noted that $V(\overline{G - x}) = V(\overline{G} - x) = V(G)\backslash x$.

Next, we shall show that $E(\overline{G - x}) = E(\overline{G} - x)$. Thus, let u, v be in $V(G)\backslash x$.

Observe that

$$uv \in E(\overline{G - x}) \Leftrightarrow uv \in E(\overline{G}) \Leftrightarrow uv \in E(\overline{G} - x);$$

and so $E(\overline{G - x}) = E(\overline{G} - x)$, as required. We thus conclude that $\overline{G - x} = \overline{G} - x$.

(ii) Yes, if x is a cut-vertex of G, then $\overline{G} - x$ is connected.

By (i), we have $\overline{G} - x = \overline{G - x}$. As x is a cut-vertex of G, $G - x$ is disconnected. Thus $\overline{G - x}$, and hence $\overline{G} - x$, is connected by Result (5). $\square$

Problem 19. (+) *(i) Show that at a gathering of any six people, some three of them are either mutual acquaintances or complete strangers to one another.*

(ii) Does the result in (i) still hold for 'five' people?

Solution. (i) We use 'graph' as a model to study the problem. Let G be a graph with $V(G) = \{a, b, c, d, e, f\}$, which represents the group of six people, such that two vertices are *adjacent* in G if and only if the two corresponding people are mutual acquaintances. Thus, two vertices are *non-adjacent* in G if and only if the two corresponding people are complete strangers to one another. Our objective is to show that either G contains a triangle or $\overline{G}$ contains a triangle.

Consider the vertex a. Either $d(a) \geq 3$ in G or $d_{\overline{G}}(a) \geq 3$. Since $\overline{\overline{G}} \cong G$, we may assume that $d(a) \geq 3$ in G. Let b, c, d be in $N(a)$. If any two in $\{b, c, d\}$ are adjacent, say b and c, then $[\{a, b, c\}]$ is a triangle in G, and we are through. Otherwise, no two in $\{b, c, d\}$ are adjacent in G, which means that $[\{a, b, c\}]$ forms a triangle in $\overline{G}$.

This completes the proof.

(ii) The result in (i) is no longer true for five vertices. Consider the cycle C_5. Both C_5 and $\overline{C_5}$ ($\cong C_5$) contain no triangles. $\square$

Problem 20. (+) *Let G be a graph of order 6. If G does not contain N_3 as an induced subgraph, what is the least possible value for $n_G(C_3)$?*

Solution. Let G be a graph of order 6 satisfying the condition

(*) *containing no N_3 as an induced subgraph.*

By Problem 19 above, G contains either a triangle or a N_3 as an induced subgraph. Since the latter cannot happen by (*), G contains at least one triangle. We claim that G contains at least 'two' triangles. Suppose on the contrary that G contains exactly one triangle, say $xyzx$, as shown below:

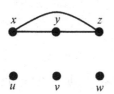

(1) Consider $\{x, u, v, w\}$. We assert that x must be adjacent to one of u, v and w. If not, then by applying (*) to $\{x, u, v\}$, $\{x, u, w\}$ and $\{x, v, w\}$, it follows that $[\{u, v, w\}]$ forms a triangle, a contradiction.

Thus, say, x is adjacent to u as shown below:

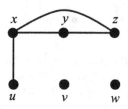

(2) Consider $\{y, u, v, w\}$. Likewise, y must be adjacent to one of u, v and w. To avoid producing another triangle, y must be adjacent to one of v and w, say v as shown below:

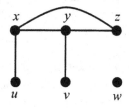

(3) Consider $\{z, u, v, w\}$. Likewise, z must be adjacent to one of u, v and w. To avoid producing another triangle, z must be adjacent to w as shown below:

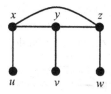

(4) Apply (*) to $\{x, v, w\}$. To avoid producing another triangle, v and w must be adjacent as shown below:

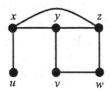

(5) Apply (*) to $\{y, u, w\}$. To avoid producing another triangle, u and w must be adjacent as shown below:

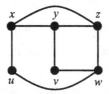

(6) Consider $\{z, u, v\}$. To avoid producing another triangle, no two in $\{z, u, v\}$ are adjacent. But then $[\{z, u, v\}] \cong N_3$, contradicting (*).

We conclude from the above discussion that G contains at least two triangles. The following graph satisfies (*) and contains exactly two triangles. Thus, the least possible value for $n_G(C_3)$ is '2'.

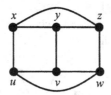

$\square$

Problem 21. *Let G be a graph with $\Delta(G) \geq r$, where r is a positive integer. Show that either G contains a triangle or $\overline{G}$ contains a K_r.*

Solution. The result is trivial if $r = 1$.

Assume that $r \geq 2$. Let w be a vertex in G such that $d(w) = \Delta(G)$ ($\geq r \geq 2$). If there are two vertices, say u and v, in $N(w)$ which are adjacent, then $[\{w, u, v\}]$ forms a triangle in G. Otherwise, $[N(w)]$ forms a complete subgraph of order at least r in $\overline{G}$, and so $\overline{G}$ contains a K_r. $\square$

Problem 22. (+) *Let G be a graph of odd order and $\delta(G) \geq 5$. Assume that G contains no N_3 as an induced subgraph. Show that G contains a K_4.*

Solution. Let G be a graph of odd order and $\delta(G) \geq 5$. By Corollary 1.2, not all vertices in G are of degree 5. Thus, there is a vertex, say w, in G with $d(w) \geq 6$. Consider $N(w)$. By Problem 19(i), $[N(w)]$ contains either a triangle or an induced N_3. The latter cannot happen by assumption. Thus, $[N(w)]$ contains a triangle. It follows that $[N(w) \cup \{w\}]$, and hence G, contains a K_4.

Remark 2.1. Indeed, by Problem 20, G contains at least two K_4's.   □

Problem 23. (∗) *Let G be a graph of order n which contains no triangles.*

(i) *Assume that $n = 9$. Show that $\overline{G}$ contains a K_4.*

(ii) *Assume that $n = 8$. Must $\overline{G}$ contain a K_4?*

Solution. (i) Let G be a graph of order 9 which contains no triangles.

    Case (1). $\Delta(G) \geq 4$. Let w be a vertex in G such that $d(w) \geq 4$ and let a, b, c, d be adjacent to w in G. As G contains no triangles, no two in $\{a, b, c, d\}$ can be adjacent in G. Thus $[\{a, b, c, d\}]$ forms a K_4 in $\overline{G}$.

    Case (2). $\Delta(G) \leq 3$. Then $\delta(\overline{G}) \geq 5$. As G contains no triangles, $\overline{G}$ contains no N_3 as an induced subgraph. Thus $\overline{G}$ contains a K_4 by the result in Problem 22.

    (ii) The conclusion in (i) is no longer true if $n = 8$. Consider the following graph G of order 8. It can be checked that G contains no triangles and $\overline{G}$ contains no K_4 as well.

G:

     □

Problem 24. (∗) *Seventeen people correspond by mail with one another - each one with all the rest. In their letters only three different topics are discussed. Each pair of correspondents deals with only one of these topics. Prove that there are at least three people who write to each other about the same topic. (IMO 1964/4)*

Solution. Before we present a proof for this problem, let us revisit Problem 19(i), which says that for any graph G of order 6, either G or $\overline{G}$ contains a triangle. By superimposing $\overline{G}$ onto G so that the same vertices are identified, we obtain a K_6. Thus the problem can equivalently be stated as: Coloring the edges of K_6 by 'blue' (for G) or 'red' (for $\overline{G}$), either there is a 'blue triangle' or a 'red triangle' in K_6.

We shall now generalize the above idea to solve the problem. Consider K_{17} in which each vertex represents a person. Color the edges in K_{17} by three colors as follows: an edge uv is colored blue (respectively, red and yellow) if u and v discuss topic I (respectively, II and III). Our aim is to show that there is a 'blue triangle', a 'red triangle' or a 'yellow triangle' in K_{17}.

Let w be a vertex in K_{17}. As the 16 edges incident with w are colored by three colors, by the Pigeonhole Principle, at least 6 of the edges are colored by one same color, say blue. Let wa, wb, wc, wd, we and wf be any six of such blue edges.

Now consider the $K_6 = [\{a, b, c, d, e, f\}]$. If one of the edges in this K_6, say ab, is colored blue, then we have a blue triangle, namely, $wabw$.

If none of the edges in this K_6 is colored blue, then all the edges in this K_6 are colored red or yellow, and so there must be a red triangle or a yellow triangle in this K_6 by the result in Problem 19(i).

The proof is thus complete.   □

Chapter 3

Bipartite Graphs and Trees

Theorem 3.1 *A graph G is bipartite if and only if it contains no odd cycles.*

Theorem 3.3 *Let G be a connected graph. Then G is a tree if and only if every two vertices in G are joined by a unique path.*

Theorem 3.4 *Let G be a connected graph of order n and size m. Then G is a tree if and only if $m = n - 1$.*

Theorem 3.5 *Let T be a tree having n_i vertices of degree i, where $i = 1, 2, \cdots, k$ with $k = \Delta(T)$. Then $n_1 = 2 + n_3 + 2n_4 + 3n_5 + \cdots + (k-2)n_k$.*

Theorem 3.6 *Let G be a graph. Then G is connected if and only if G contains a spanning tree.*

Corollary 3.7 *If G is a connected graph of order n and size m, then*

$$m \geq n - 1.$$

Result (1). Let G be a bipartite graph with a bipartition (X, Y). Then

$$\sum_{x \in X} d(x) = e(G) = \sum_{y \in Y} d(y).$$

Exercise 3.1

Problem 1. *For each of the following cases, construct all desired connected bipartite graphs H of order n:*

 (i) $2 \leq n \leq 4$;

 (ii) $n = 5$ and H contains no cycles;

(iii) $n = 5$ and H contains a cycle;

(iv) $n = 6$ and H contains a C_6;

 (v) $n = 8$, H is 3-regular and contains a C_8.

Solution.   (i)

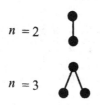

(ii)

$n = 5$

(iii)

$n = 5$

(iv) $n = 6$.

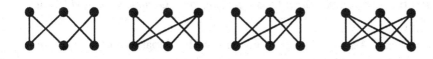

(v) $n = 8$.

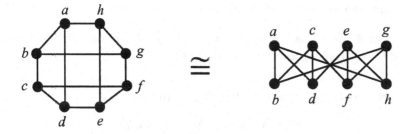

$\cong$

□

Problem 2. *Let G be a connected bipartite graph. Then G has a bipartition (X, Y). Is $\{X, Y\}$ always unique? What if G is disconnected?*

Solution. If G is connected, then $\{X, Y\}$ is unique. Indeed, if we fix a vertex w in G and, say, w is in X, then

$$X = \{v \in V(G) \mid d(w, v) \text{ is even}\}$$

and

$$Y = \{v \in V(G) \mid d(w, v) \text{ is odd}\}.$$

If G is disconnected, then $\{X, Y\}$ need not be unique. An example is given below:

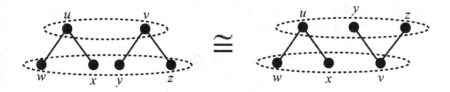

$\cong$

□

Problem 3. *Show that if G is bipartite, then G contains no odd cycles.*

Solution. Assume that G is bipartite with a bipartition (X, Y), and suppose on the contrary that G contains an odd cycle $v_1 v_2 \cdots v_{2k+1} v_1$. We may assume that $v_1 \in X$. As G is bipartite, it follows that $v_2 \in Y$, $v_3 \in X, \cdots, v_{2k} \in Y$ and $v_{2k+1} \in X$. Thus $\{v_1, v_{2k+1}\} \subseteq X$, but v_1 and v_{2k+1} are adjacent in G, a contradiction. □

Problem 4. *Let G be a bipartite graph with a bipartition (X, Y). Show that if G is k-regular, where $k \geq 1$, then $|X| = |Y|$.*

Solution. As G is k-regular, by Result (1), we have

$$k|X| = \sum_{x \in X} d(x) = \sum_{y \in Y} d(y) = k|Y|.$$

As $k \neq 0$, it follows that $|X| = |Y|$. □

Problem 5. *Construct all non-isomorphic graphs of order 8 and size 10 that are bipartite and contain a C_8.*

Solution. By assumption, G contains a C_8:

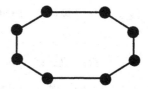

As $e(G) = 10$, we need to add 2 more edges, say 'e' and 'f'. Note that G contains no odd cycles.

Case (1). e and f are adjacent (i.e., they are incident with a common vertex). Then G is the following graph:

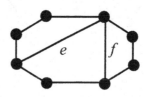

Case (2). e and f are not adjacent. Then G is isomorphic to one of the following:

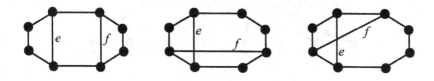

Note. The reader should verify that the three graphs in Case (2) are not isomorphic.

Problem 6. *Let G be a bipartite graph of order n with a bipartition (X, Y). Assume that G contains a cycle C_n. What is the relation between $|X|$ and $|Y|$?*

Solution. Let G be a bipartite graph of order n with a bipartition (X, Y). Suppose that G contains a C_n (thus, $n \geq 3$). As G contains no odd cycles, $n = 2k$ for some positive integer k. Write C as $v_1 v_2 \cdots v_{2k} v_1$ and assume that $v_1 \in X$. Clearly, the mapping f defined by $f(v_1) = v_2, f(v_3) = v_4, \cdots, f(v_{2k-1}) = v_{2k}$ is a one-to-one mapping from X onto Y. Thus $|X| = |Y|$. □

Note: As an extension of Problems 4 and 6, if G is a bipartite graph with a bipartition (X, Y) such that it contains a k-regular spanning subgraph, then $|X| = |Y|$.

Problem 7. *Does there exist a bipartite graph with degree sequence* $(5, 5, 5, 4, 4, 3, 3, 3, 1, 1, 1, 1)$? *Justify your answer.*

Solution. Yes, such a bipartite graph is given below:

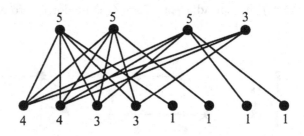

Problem 8. (+) *Show that there does not exist a bipartite graph with degree sequence* $(6, 6, \cdots, 6, 5, 3, 3, \cdots, 3)$.

Solution. Suppose on the contrary that there exists a bipartite graph G with a bipartition (X, Y) and degree sequence $(6, 6, \cdots, 6, 5, 3, 3, \cdots, 3)$.

Let w be a vertex in G such that $d(w) = 5$. We may assume that $w \in X$. By Result (1), we have

$$\sum_{x \in X} d(x) = \sum_{y \in Y} d(y). \qquad (*)$$

Now observe that, as $w \in X$, the sum $\sum_{y \in Y} d(y)$ is divisible by 3 while $\sum_{x \in X} d(x)$ is not so. This, however, contradicts (*).

Thus, such a bipartite graph G does not exist. □

Problem 9. (+) *At a party, assume that no boy dances with every girl but each girl dances with at least one boy. Prove that there are two couples b, g and b', g' which dance, whereas b does not dance with g' nor does g dance with b'. (Putnam Exam (1965))*

Solution. Let X be the set of boys and Y the set of girls at a party. Let G be a bipartite graph in which b ($\in X$) and g ($\in Y$) are joined by an edge if and only if b dances with g. Our aim is to show that there exist b, b' in X and g, g' in Y such that the adjacency relations shown in the following diagram holds:

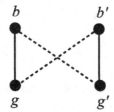

where "●······●" indicates the non-adjacency of the two vertices.

Let b be a vertex in X such that $d(b) = \max\{d(x) \mid x \in X\}$. Clearly, $d(b) \geq 1$; i.e. $N(b)$ is non-empty. As no boy dances with every girl, there exists a vertex, say g', in $Y\backslash N(b)$. Since each girl dances with at least one boy, there exists a vertex, say b', in X such that b' and g' are adjacent. Clearly, $b' \neq b$. Now, if b' is adjacent to every vertex in $N(b)$, then $d(b') \geq d(b) + 1$, contradicting the maximality of $d(b)$ in X. Thus, there exists a vertex, say g, in $N(b)$ which is not adjacent to b'. It follows that the four vertices chosen, namely b, b', g and g', are the desired ones (see the diagram below).

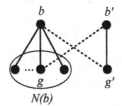

N(b)

□

Problem 10. (+) *Let G be a bipartite graph with a bipartition (X, Y). Assume that $e(G) = v(G)$, and that $d(x) \leq 5$ for each x in X. Show that $|Y| \leq 4|X|$.*

Solution. By Result (1) and assumption, we have:

$$\sum_{x \in X} d(x) = e(G) = v(G).$$

Since $d(x) \leq 5$ for each x in X and $V(G) = X \cup Y$, we have:

$$|X| + |Y| = v(G) = \sum_{x \in X} d(x) \leq 5|X|.$$

Thus, $|Y| \leq 4|X|$, as required.

□

Problem 11. (+) *Let H be a bipartite graph with a bipartition (X, Y). Assume that $e(H) \leq 2v(H)$, and $d(x) \geq 3$ for each x in X. Show that $|X| \leq 2|Y|$.*

 Construct one such graph H with $|X| = 2|Y|$.

Solution. By Result (1) and assumption, we have:

$$\sum_{x \in X} d(x) = e(H) \leq 2v(H).$$

As $d(x) \geq 3$ for each x in X and $V(H) = X \cup Y$, we have:

$$2(|X| + |Y|) = 2v(H) \geq \sum_{x \in X} d(x) \geq 3|X|.$$

It follows that $|X| \leq 2|Y|$.

 An example of H with $|X| = 2|Y|$ is shown below.

$K(6,3)$:

Note that in this example, $v(H) = 9$ and $e(H) = 18$.   □

Problem 12. *Let G be a bipartite graph of order $2k$, where k is a positive integer. What is the maximum size of G? Find all such bipartite graphs with maximum size.*

Solution. We claim that the maximum size of G is achieved if and only if $G = K(k, k)$, and in this case, $e(G) = k^2$. A proof is shown below.

 Let G be any bipartite graph of order $2k$ with a bipartition (X, Y), where $|X| = x$ and $|Y| = 2k - x$. Then $e(G) \leq x(2k - x)$, and the equality holds if and only if G is a complete bipartite graph, i.e., $G \cong K(x, 2k - x)$. Note that

$$x(2k - x) = -(x - k)^2 + k^2 \leq k^2,$$

and the equality holds if and only if $x = k$. Our claim thus follows. □

Problem 13. *Let G be a bipartite graph of order $2k + 1$, where k is a positive integer. What is the maximum size of G? Find all such bipartite graphs with maximum size.*

Solution. The maximum size of G is achieved if and only if $G = K(k, k+1)$, and in this case, $e(G) = k(k + 1)$. It can similarly be proved as shown in the solution of Problem 12. □

Problem 14. *Find, in terms of p and q, the number of C_4 in $K(p,q)$, where $2 \leq p \leq q$.*

Solution. The number of C_4 is $\binom{p}{2}\binom{q}{2}$. □

Problem 15. *Find, in terms of p and q, the number of C_6 in $K(p,q)$, where $3 \leq p \leq q$.*

Solution. The number of C_6 is $3 \times 2 \times \binom{p}{3}\binom{q}{3} = 6\binom{p}{3}\binom{q}{3}$. □

Problem 16. *Let H be a graph obtained from $K(p,q)$, $2 \leq p \leq q$, by adding a new edge.*

(i) *Is H bipartite?*

(ii) *What is the largest number of triangles that H could contain?*

(iii) *(+) What is the largest number of C_5 that H could contain?*

Solution. Let (X, Y) be the bipartition of $K(p,q)$, where $|X| = p$. Let e be the new edge added to form H. Then e can only join two vertices in X or two vertices in Y.

(i) No, H is no longer bipartite as H contains a triangle.

(ii) When e joins two vertices in X, H has the largest number of triangles, which is q.

(iii) When e joins two vertices in X, the number of C_5's in H (see the diagram below) is $2\binom{q}{2}(p - 2)$; and while e joins two vertices in Y, the number of C_5's in H is $2\binom{p}{2}(q - 2)$. As $p \leq q$, solving the following inequality:

$$2\binom{q}{2}(p - 2) < 2\binom{p}{2}(q - 2)$$

gives the solution that $p = 2$ and $q \geq 3$. Thus the largest number of C_5's that H could contain is

$$\begin{cases} 2(q - 2), & \text{if } p = 2; \\ 2\binom{q}{2}(p - 2), & \text{if } q \geq p \geq 3. \end{cases}$$

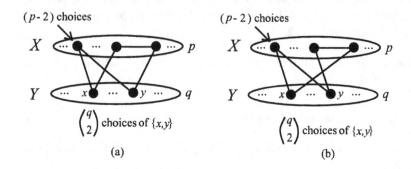

(a)   (b)

Problem 17. *What is the largest cycle in $K(p,q)$, where $2 \leq p \leq q$?*

Solution. The largest cycle in $K(p,q)$ is a cycle of order $2p$, where $2 \leq p \leq q$.

Problem 18. *What can be said about the complement of $K(p,q)$?*

Solution. The complement of $K(p,q)$ is the disjoint union of a K_p and a K_q.

Problem 19. *Let G be a bipartite graph.*

(i) *Is $\overline{G}$ also bipartite?*

(ii) *Is $\overline{G}$ always connected?*

(iii) *What conditions should be imposed on G so that $\overline{G}$ is connected?*

Solution. Let G be a bipartite graph with a bipartition (X, Y), where $|X| = p$ and $|Y| = q$.

(i) No, $\overline{G}$ may not be bipartite. For instance, $\overline{K_{1,3}}$ contains a triangle.

(ii) No, $\overline{G}$ may not be connected. For instance, $\overline{K_{1,3}}$ is disconnected.

(iii) $\overline{G}$ is connected if and only if G is not a complete bipartite graph. We justify it as follows.

($\Rightarrow$) If G is a complete bipartite graph, then (see Problem 18 above) $\overline{G}$ has two components, namely a K_p and a K_q, a contradiction.

($\Leftarrow$) If G is not a complete bipartite graph, then x and y are not adjacent for some x in X and some y in Y. This implies that $\overline{G}$ contains the following connected graph as a spanning subgraph.

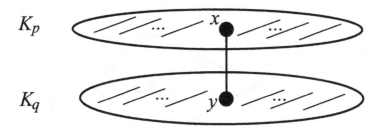

Thus, $\overline{G}$ is connected.   □

Problem 20. (+) *A connected graph G has the following property:*

For each pair of distinct vertices u and v, either all $u - v$ paths are of even length or all $u - v$ paths are of odd length.

What can be said about G? Justify your answer.

Solution. The graph G must be bipartite. We prove it by contradiction as follows.

Suppose on the contrary that G is not bipartite. Then, by Theorem 3.1, G contains an odd cycle C. Let u and v be any two vertices in C. Then, as C is odd, the two different $u - v$ paths along C have lengths of different parity (one odd and one even; see the diagram below), a contradiction.

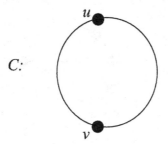

Note. Every bipartite graph has the property described in the problem. □

Problem 21. (∗) *Let G be a graph. A cycle C in G is said to be* **induced** *if C is induced by $V(C)$.*

(i) *Consider the following graph H. Which cycles in H are induced cycles?*

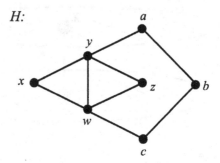

(ii) *Show that G is bipartite if and only if G contains no induced cycles of odd order.*

Solution. (i) There are three induced cycles, namely, $xywx, yzwy$ and $abcwya$.

(ii) ($\Rightarrow$) Suppose on the contrary that G contains an induced cycle C of odd order. Then C is itself an odd cycle in G, and so G is not bipartite by Theorem 3.1, a contradiction.

($\Leftarrow$) Suppose on the contrary that G is not bipartite. Then, by Theorem 3.1, G contains an odd cycle C. If C is induced, then we are through; otherwise, there are two vertices, say u and v, in C, which are not adjacent along C but are adjacent in G (see the diagram below).

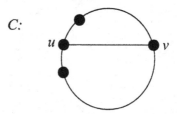

As C is odd, one of the $u - v$ paths along C forms, together with the edge uv, a smaller odd cycle C'. If C' is induced, then we are through; otherwise, the above argument can be repeated in a finite number of steps to eventually reveal an induced cycle of odd order in G.   □

Problem 22. (+) *A graph H has the property that each edge in H is incident with an even vertex and an odd vertex. What can be said about H? Construct one such H.*

Solution. The graph H must be bipartite. We prove it by contradiction as follows.

Suppose on the contrary that H is not bipartite. Then, by Theorem 3.1, H contains an odd cycle $C : v_1 v_2 \cdots v_{2k+1} v_1$. We may assume that v_1 is odd. But then, by assumption, v_2 is even, which, in turn, implies eventually that v_{2k+1} is odd. This, however, contradicts the assumption as these two odd vertices v_1 and v_{2k+1} are joined an edge.

The graphs $K(p, q)$, where p and q are of different parity, are examples.
Note. Not every bipartite graph has the property described in the problem.
□

Problem 23. *Let G be a bipartite graph of order 7 such that every vertex in G is contained in a cycle.*

(i) *Construct one such G.*

(ii) *Must G be connected?*

(iii) *What is the least possible value of e(G)?*

(iv) *Construct all non-isomorphic graphs G which have their e(G) attaining the least possible value obtained in (iii).*

Solution. (i) Take $K(2, 5)$ or $K(3, 4)$.

(ii) Yes, G must be connected. The justification is as follows. If G is disconnected, then, as $v(G) \geq 7$, one of its components, say H, contains at most three vertices. But then no vertex in H is contained in a cycle as C_4 is the smallest possible cycle in G.

(iii) Claim. $e(G) \geq 8$.

Let (X, Y) be the bipartition of G. As $v(G) \geq 7$, either $|X| \geq 4$ or $|Y| \geq 4$, say the latter. By assumption, every vertex is contained in a cycle. It follows that $d(v) \geq 2$ for each vertex v in G. Thus, by Result (1), we have:

$$e(G) = \sum_{y \in Y} d(y) \geq 2|Y| \geq 8.$$

(iv) All such extremal graphs are shown below:

 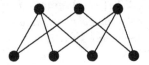

□

Problem 24. *Let G be a connected bipartite graph of order $p + q$ and size pq, where $1 \le p \le q$. Is it true that $G \cong K(p, q)$?*

Solution. No, take, for instance, $p = 2$ and $q = 4$, and consider the following connected bipartite graph G of order $6(= 2 + 4)$ and size 8 ($= 2 \times 4$).

G:

Clearly, G is not isomorphic to $K(2, 4)$.   □

Problem 25. (+) *Let G be a bipartite graph of order $p + q$ and size pq, where $2 \le p \le q$, and with $\delta(G) \ge 1$. Show that $G \cong K(p, q)$ if and only if every two edges in G are contained in a common C_4.*

Solution. ($\Rightarrow$) If $G \cong K(p, q)$, where $2 \le p \le q$, then it is clear that every two edges in G are contained in a common C_4.

($\Leftarrow$) Let (X, Y) be a bipartition of G.

Claim. G is a complete bipartite graph.

If not, then there exist u in X and v in Y which are not adjacent in G. As $\delta(G) \ge 1$, assume that u is adjacent to y in Y and v is adjacent to x in X (see the diagram below).

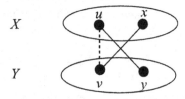

By assumption, the edges uy and vx are contained in a common C_4. This, however, implies that uv is an edge, a contradiction. Thus, G is a complete bipartite graph with bipartition (X, Y).

We shall now show that $G \cong K(p, q)$.

Let $|X| = k$, where $2 \le k \le p + q - 2$. Then $|Y| = p + q - k$ and $pq = e(G) = k(p + q - k)$. The latter implies that $(k - p)(k - q) = 0$. Thus either $k = p$ or $k = q$.

Now, if $k = p$, then $|Y| = q$, and we have $G \cong K(p, q)$.

If $k = q$, then $|X| = q$ and $|Y| = p$; that is, $G \cong K(q, p) \cong K(p, q)$. $\quad\square$

Exercise 3.2

Problem 1. *Draw all non-isomorphic trees of order n, where $2 \le n \le 6$.*

Solution.

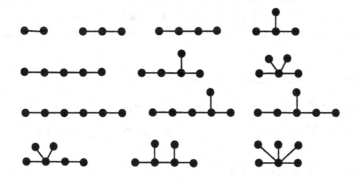

$\square$

Problem 2. *Let G be a graph of order n and size $n - 1$, where $n \ge 4$. Must G be a tree?*

Solution. No, an example ($n = 4$) is shown below:

$\square$

Problem 3. *Let T be a tree of order $n \geq 2$. Show that T has exactly two end-vertices if and only if T is a path, i.e., $T \cong P_n$.*

Solution. Let T be a tree of order $n \geq 2$.

If T is a path: $v_1 v_2 \cdots v_n$, then v_1 and v_n are the only end-vertices.

If T is not a path, then, by definition, T contains a vertex, say w, such that $d(w) \geq 3$. Let x, y and z be in $N(w)$ (see the diagram below).

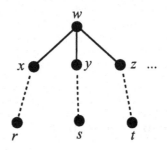

Traversing from w through x (respectively, y and z) via edges in T as far as possible, as T contains no cycles, we shall terminate at a vertex, say r (respectively, s and t). Clearly, r, s and t are three end-vertices in T, a contradiction.   □

Problem 4. *Let T be a tree of order $n \geq 3$.*

   (i) *What is T if $d(x, y) \leq 2$ for any two vertices x, y in T?*

  (ii) *What is T if $d(u, v) = n - 1$ for some vertices u, v in T?*

Solution.   (i) T is the star $K(1, n - 1)$.

         (ii) T is the path P_n.   □

Problem 5. *Find all trees T of order $n \geq 2$ such that $\overline{T}$ is a tree. Is there any tree of order $n \geq 2$ which is self-complementary?*

Solution.   Let T be a tree of order $n \geq 2$. Then $e(T) + e(\overline{T}) = n(n-1)/2$. As both T and $\overline{T}$ are trees, $e(T) = e(\overline{T}) = n - 1$. Thus $2(n-1) = n(n-1)/2$. As $n \geq 2$, it follows that $n = 4$.

It can be checked that the path of order 4 (•——•——•——•) is the only tree T such that $\overline{T}$ is a tree. Note that T is also self-complementary.   □

Problem 6. *A connected graph is said to be* **unicyclic** *if it contains one and only one cycle as a subgraph.*

(i) *Is every cycle unicyclic?*

(ii) *Construct two unicyclic graphs of order 8 which are not C_8.*

(iii) *How many edges are there in each of your graphs in (ii)?*

Solution. (i) Yes, every cycle is unicyclic.
(ii) Two such graphs are shown below:

(iii) Each graph is of size 8 (same as the order). □

Problem 7. *Let G be a unicyclic graph.*

(i) *What is the relation between $e(G)$ and $v(G)$? Justify your answer.*

(ii) *Show that there exist at least three edges e in G such that $G - e$ is a tree.*

Solution. Let G be a unicyclic graph.
(i) Then $e(G) = v(G)$.
Let f be any edge contained in the only cycle in G. Observe that $G - f$ is still connected (see Problem 16(iii) in Exercise 2.3) and it contains no cycle. Thus, $G - f$ is a tree, and by Theorem 3.4, $e(G - f) = v(G - f) - 1$. It follows that $e(G) = e(G - f) + 1 = v(G - f) = v(G)$, as asserted.
(ii) Let C_k be the cycle in G. As shown in (i), the deletion of any edge in C_k results in a tree. The result now follows as $k \geq 3$. □

Problem 8. (+) *Let G be a unicyclic graph and let n_1 denote the number of end-vertices in G. Find an expression for n_1 similar to that in Theorem 3.5.*

Solution. Let G be a unicyclic graph of order n and let n_i denote the number of vertices of degree i in G, $i = 1, 2, \cdots, k(= \Delta(G))$. Then, by

Problem 7(i), $e(G) = v(G) = n$, and by Theorem 1.1,

$$\sum_{v \in V(G)} d(v) = 2e(G) = 2n.$$

But

$$\sum_{v \in V(G)} d(v) = n_1 + 2n_2 + \cdots + kn_k.$$

Thus

$$n_1 + 2n_2 + \cdots + kn_k = 2n = 2(n_1 + n_2 + \cdots + n_k).$$

and we have

$$n_1 = n_3 + 2n_4 + \cdots + (k-2)n_k.$$

$\square$

Problem 9. *Let G be a connected graph. Show that G is a tree if and only if every edge in G is a bridge. (See Problem 16 in Exercise 2.3.)*

Solution. Let G be a connected graph.

$(\Rightarrow)$ Assume that G is a tree. Suppose that G contains an edge, say f, which is not a bridge. Then, by the result of Problem 16 in Exercise 2.3, f is contained in a cycle in G, which is impossible as G contains no cycles.

$(\Leftarrow)$ Assume that every edge in G is a bridge. Suppose that G is not a tree. Then G contains a cycle C. However, by the result of Problem 16 in Exercise 2.3, each edge in C is not a bridge, a contradiction. $\square$

Problem 10. $(*)$ *Let T be a tree of order k. Show that if G is a graph with $\delta(G) \geq k - 1$, then T is isomorphic to some subgraph of G.*

Solution. The proof is by induction on k. For $k = 1, 2$, the result is trivial.

Assume that the result is true when $k = n$. We now consider the case when $k = n + 1$. Thus, let T be a tree of order $n + 1$ and G be a graph with $\delta(G) \geq n$. We shall show that T is isomorphic to some subgraph of G. Let w be an end-vertex of T and suppose that w is adjacent to u in T (see the diagram below). Write $T' = T - w$. Since $v(T') = n$ and $\delta(G) \geq n > n - 1$, by the induction hypothesis, T' is isomorphic to some subgraph, say T'', of G.

Let u' be the image of u in G under an isomorphism. Observe that $d(u') \geq \delta(G) \geq n$ in G and $v(T'') = n$. Thus, u' is adjacent to a vertex in G, say v, which is not in T''. Clearly, the subgraph of G, which consists of T'' and the edge $u'v$, is isomorphic to T. The proof is thus complete. $\square$

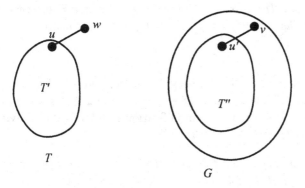

Problem 11. Let T be a tree of order 15 such that $1 \le d(v) \le 4$ for each vertex v in T. Suppose that T contains exactly 9 end-vertices and exactly 3 vertices of degree 4. How many vertices of degree 3 does T have? Justify your answer. Construct one such tree T.

Solution. For $i = 1, 2, 3, 4$, let n_i denote the number of vertices of degree i in T. Then $n_1 = 9$, $n_4 = 3$ and so

$$n_2 + n_3 = 15 - 9 - 3 = 3. \qquad (1)$$

By Theorem 1.1 and Theorem 3.4, we have:

$$n_1 + 2n_2 + 3n_3 + 4n_4 = 2e(T) = 2(v(T) - 1) = 28,$$

or

$$2n_2 + 3n_3 = 28 - 9 - 4 \times 3 = 7. \qquad (2)$$

Solving (1) and (2) yields $(n_2, n_3) = (2, 1)$.

An example of T is shown below:

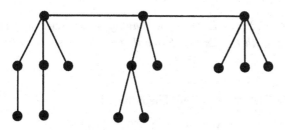

$\square$

Problem 12. *The degrees of the vertices of a tree T of order 18 are $1, 2$ and 5. If T has exactly 4 vertices of degree 2, how many end-vertices does T have?*

Solution. Let x and y be, respectively, the number of vertices of degree 1 and 5 in T. Then $x + y = 18 - 4 = 14$, and $x + 2 \times 4 + 5y = 2e(T) = 2(v(T) - 1) = 34$ or, upon simplification $x + 5y = 26$. Solving the two equations yields $(x, y) = (11, 3)$. □

Problem 13. *Let T be a tree and let n_i be the number of vertices of degree i in T. Which of the following statements is/are true?*

(i) *If T is not a path, then $n_1 \geq n_2$.*

(ii) *If $n_2 = 0$, then T has more end-vertices than other vertices.*

Solution. (i) False. An example is shown below:

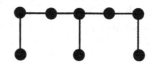

The above tree is not a path; and while $n_1 = 3, n_2 = 4$.

(ii) The statement is true. Indeed, by Theorem 3.5,

$$
\begin{aligned}
n_1 &= 2 + n_3 + 2n_4 + \cdots + (k-2)n_k & (k = \Delta(G)) \\
&> n_2 + n_3 + \cdots + n_k & (n_2 = 0) \\
&= n - n_1.
\end{aligned}
$$

□

Problem 14. (+) *Let T be a tree having n_i vertices of degree i, where $i = 1, 2, \cdots, k$, with $k = \Delta(T)$. Show that $n_1 = 2 + n_3 + 2n_4 + 3n_5 + \cdots + (k-2)n_k$.*

Solution. By Theorem 1.1 and Theorem 3.4,

$$
\sum_{v \in V(T)} d(v) = 2e(T) = 2(n-1),
$$

where $n = v(T)$.

But $\sum\limits_{v\in V(T)} d(v) = n_1 + 2n_2 + \cdots + kn_k$. Thus

$$n_1 + 2n_2 + \cdots + kn_k = 2(n-1) = 2(n_1 + n_2 + \cdots + n_k) - 2,$$

and we have $n_1 = 2 + n_3 + 2n_4 + 3n_5 + \cdots + (k-2)n_k$, as was to be shown.
□

Problem 15. (+) *Let T be a tree of order n. Show that the vertices in T can always be named as $x_1, x_2, \cdots, x_n$ so that every x_i has one and only one neighbour in $\{x_1, x_2, \cdots, x_{i-1}\}$, for $i = 2, 3, \cdots, n$.*

Solution. We apply the procedure introduced in the solution of Problem 24 in Exercise 2.3 to name the n vertices in T as $x_1, x_2, \cdots, x_n$. Recall that $[\{x_1, x_2, \cdots, x_i\}]$ is connected for each $i = 1, 2, \cdots, n$. It is clear from the procedure that every x_i has a neighbour in $\{x_1, x_2, \cdots, x_{i-1}\}$. Suppose x_i has two neighbours, say x_j and x_k, in $\{x_1, x_2, \cdots, x_{i-1}\}$. As $[\{x_1, x_2, \cdots, x_{i-1}\}]$ is connected, it contains a $x_j - x_k$ path. But then this path, together with the edges $x_i x_j$ and $x_i x_k$, forms a cycle in T, a contradiction.
□

Problem 16. *Let G be a graph of order n with degree sequence $(d_1, d_2, \cdots, d_n)$. Show that if G is a tree, then*

$$\sum_{i=1}^{n} d_i = 2(n-1).$$

Is the converse true?

Solution. If G is a tree, then, by Theorem 3.4, $e(G) = n - 1$; and we have, by Theorem 1.1,

$$\sum_{i=1}^{n} d_i = 2e(G) = 2(n-1).$$

The converse is not true in general. An example ($n = 5$) is shown below:

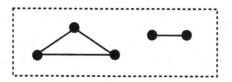

□

Problem 17. (∗) *Show that every sequence $(d_1, d_2, \cdots, d_n)$ of positive integers with*

$$\sum_{i=1}^{n} d_i = 2(n-1)$$

is a degree sequence of a tree.

Solution.   We shall prove the statement by induction on $n(\geq 2)$.

For $n = 2$, as $d_1 + d_2 = 2$, $d_1 = d_2 = 1$; and the sequence $(1, 1)$ is the degree sequence of the path of order 2.

Assume that the statement is true for all sequences of length $n - 1$, where $n \geq 3$.

Consider now the sequence $(d_1, d_2, \cdots, d_n)$ with

$$\sum_{i=1}^{n} d_i = 2(n-1).$$

We may assume that $d_1 \geq d_2 \geq \cdots \geq d_n$.

Clearly, $d_n = 1$; otherwise, $d_1 \geq d_2 \geq \cdots \geq d_n \geq 2$, and we have

$$\sum_{i=1}^{n} d_i \geq 2n,$$

a contradiction.

As $n \geq 3$, let k be the largest index in $\{1, 2, \cdots, n-1\}$ such that $d_k \geq 2$. Let $d'_k = d_k - 1$. Then

$$(d_1, d_2, \cdots, d_{k-1}, d'_k, \cdots, d_{n-1}) \qquad (1)$$

is a sequence of positive integers of length $n - 1$ such that

$$d_1 + d_2 + \cdots + d_{k-1} + d'_k + \cdots + d_{n-1}$$
$$= 2(n-1) - 2 \qquad (d_n = 1 \text{ and } d'_k = d_k - 1)$$
$$= 2((n-1) - 1).$$

By the induction hypothesis, the sequence (1) is the degree sequence of some tree T of order $n - 1$.

Let v be a vertex in T of degree d'_k. Construct a tree T^* of order n by adding to T a new vertex w and joining w to v (thus $d(w) = 1$ and $d(v) = d_k$ in T^*). Clearly, $(d_1, d_2, \cdots, d_n)$ is the degree sequence of T^*.

The proof is thus complete.   □

Problem 18. *A* **forest** *is a graph which contains no cycle as a subgraph.*

(i) *Is it true that every tree is a forest?*

(ii) *Is it true that every forest is a tree?*

(iii) *Is it true that every connected component of a forest is a tree?*

(iv) (+) *Let F be a forest. Find a relation linking $v(F), e(F)$ and $c(F)$, and prove your result.*

Solution. (i) True.

(ii) False. A forest may not be connected.

(iii) True.

(iv) Let $c(F) = k$ and assume that $F_1, F_2, \cdots, F_k$ are the components of F. As each F_i is a tree, we have, by Theorem 3.4,

$$e(F) = e(F_1) + e(F_2) + \cdots + e(F_k)$$
$$= (v(F_1) - 1) + (v(F_2) - 1) + \cdots + (v(F_k) - 1)$$
$$= v(F) - k;$$

That is, $v(F) = e(F) + c(F)$.   □

Problem 19. (+) *Let G be a graph of order n and size $n - 1$. Prove that G is connected if and only if G contains no cycles.*

Solution. Let G be a graph of order n and size $n - 1$.

($\Rightarrow$) Assume that G is connected. Since $e(G) = v(G) - 1$, by Theorem 3.4, G is a tree, and so contains no cycles.

($\Leftarrow$) Assume that G contains no cycles. Then G is a forest (see Problem 18). By the result of Problem 18(iii), $c(G) = v(G) - e(G)$. Thus, by the given assumption, $c(G) = n - (n-1) = 1$, which means that G is connected. □

(∗) Exercise 3.3

Problem 1. *Find all spanning trees of the following graph.*

Solution. The graph has 8 spanning trees which are shown below:

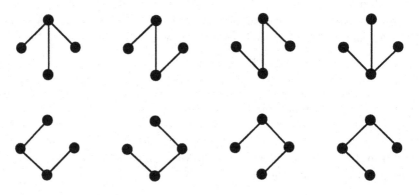

□

Problem 2. *Let H be a graph of order 1000 and size 998. Can H be connected? Why?*

Solution. No, H cannot be connected. For if H is connected, then by Corollary 3.7, $998 = e(H) \geq v(H) - 1 = 1000 - 1$, which is impossible. □

Problem 3. *Let G be a connected graph and e be a bridge in G. Must e be contained in any spanning tree of G? Why?*

Solution. Let e be a bridge in a connected graph G. Then e must be contained in any spanning tree of G. For if e is not contained in a spanning tree, say T, of G, then T is also a spanning tree of $G - e$, and so $G - e$ is connected by Theorem 3.6. This, however, contradicts the fact that e is a bridge in G. □

Problem 4. *Let G be a unicyclic graph which contains C_k as a subgraph, where $k \geq 3$. How many spanning trees does G have?*

Solution. A spanning tree of G can be obtained by and only by deleting an edge in C_k from G. Thus G has exactly k spanning trees. □

Problem 5. (+) *Prove that every graph of order n and size $n - r$ has at least r components.*

Solution. Let G be a graph of order n and size $n - r$. Assume that G has k components, say $G_1, G_2, \cdots, G_k$. We shall show that $k \geq r$.

Indeed, we have:

$$n - r = e(G)$$
$$= e(G_1) + e(G_2) + \cdots + e(G_k)$$
$$\geq (v(G_1) - 1) + (v(G_2) - 1) + \cdots + (v(G_k) - 1)$$
$$\text{(by Corollary 3.7)}$$
$$= n - k;$$

that is, $k \geq r$, as required. □

Problem 6. *Let G be a connected bipartite graph with bipartition (X, Y). Assume that $d(x) \leq 7$ for each x in X. Show that*

$$|Y| \leq 6|X| + 1.$$

For each $|X| = 1, 2, \cdots$, construct one such bipartite graph G with $|Y| = 6|X| + 1$.

Solution. Let G be a connected bipartite graph with bipartition (X, Y).

By Result (1) in Section 3.1,

$$e(G) = \sum_{x \in X} d(x).$$

As $d(x) \leq 7$ for each x in X, $e(G) \leq 7|X|$.

On the other hand, as G is connected, by Corollary 3.7,

$$e(G) \geq v(G) - 1 = |X| + |Y| - 1.$$

Combining the above two inequalities, we have:

$$|X| + |Y| - 1 \leq e(G) \leq 7|X|;$$

that is, $|Y| \leq 6|X| + 1$.

An example of such a connected bipartite graph with $|Y| = 6|X| + 1$ is shown below:

Problem 7. *Let G be a connected bipartite graph with bipartition (X, Y). Assume that G is not a tree and $d(x) \leq 4$ for each x in X. Find the best upper bound for $|Y|$, in terms of $|X|$. Justify your answer.*

Solution. Let G be a connected bipartite graph with bipartition (X, Y). By Result (1) in Section 3.1,

$$e(G) = \sum_{x \in X} d(x).$$

As $d(x) \leq 4$ for each x in X,

$$e(G) = \sum_{x \in X} d(x) \leq 4|X|.$$

On the other hand, as G is connected but not a tree, $e(G) > v(G) - 1$ by Corollary 3.7 and Theorem 3.4.

Combining the above two inequalities, we have:

$$|X| + |Y| = v(G) \leq e(G) \leq 4|X|;$$

that is, $|Y| \leq 3|X|$. The following example shows that $|Y| = 3|X|$ holds when $|X| = 2$.

Note. The reader is encouraged to construct such G for which $|Y| = 3|X|$ holds for $|X| = 3, 4, \cdots$. □

Chapter 4

Vertex-colourings of Graphs

Theorem 4.1 *For any graph G, $\chi(G) \leq \Delta(G) + 1$.*

Theorem 4.2 *Let G be a connected graph which is neither an odd cycle nor a complete graph. Then $\chi(G) \leq \Delta(G)$.*

Result (1). Let G be a graph of order n. Then $\chi(G) = 1$ if and only if $G \cong N_n$.

Result (2). Let G be a graph of order n. Then $\chi(G) = n$ if and only if $G \cong K_n$.

Result (3). Let G be a graph with at least one edge. Then $\chi(G) = 2$ if and only if G is bipartite.

Result (4). Let G be a graph which contains an odd cycle as a subgraph. Then $\chi(G) \geq 3$.

Result (5). Let G be a graph and let p be any positive integer such that G contains a K_p as a subgraph. Then $\chi(G) \geq p$.

Exercise 4.2

Problem 1.   *Consider the following map:*

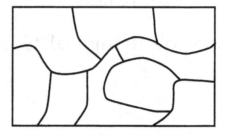

(i) *Colour the regions with no more than four colours in such a way that each region is coloured by one colour, and adjacent regions are coloured by different colours.*

(ii) *Construct a graph G modeling the above situation as shown in Figure 4.3.*

(iii) *Does G contain a K_4 as a subgraph?*

(iv) *Does G contain a K_5 as a subgraph?*

(v) *Is G 3-colourable? Why?*

(vi) *Is G 4-colourable?*

(vii) *What is the value of $\chi(G)$?*

Solution.

(i) The map is coloured with four colours 1, 2, 3 and 4.

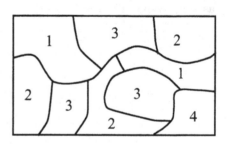

(ii) The graph G below models the map.

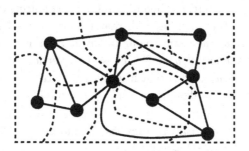

(iii) Yes, G contains a K_4 as a subgraph.

(iv) No, G does not contain K_5 as a subgraph.

(v) G is not 3-colourable because it contains a K_4 as a subgraph.

(vi) Yes, G is 4-colourable since the map colouring in (i) can be translated into a 4-colouring of G as shown below.

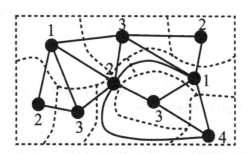

(vii) $\chi(G) = 4$. $\qquad\qquad\qquad\qquad\qquad\qquad\qquad\qquad\qquad$ □

Problem 2. *Let p and q be integers such that $1 \le p \le q$. Explain by definition why a p-colouring of a graph is also a q-colouring of the graph.*

Solution. Let θ be a p-colouring of a graph G. By definition, the number of colours x used by θ to colour the vertices of G is at most p, i.e. $x \le p$. Since $p \le q$, we have $x \le q$. Thus, the number of colours x used by θ to colour the vertices of G is at most q as well and so θ is also a q-colouring of G. $\qquad\qquad$ □

Problem 3. *Prove that if H is a subgraph of a graph G, then $\chi(H) \le \chi(G)$.*

Solution. Let $k = \chi(G)$. Then G admits a k-colouring θ. Let $\alpha : V(H) \to \{1, 2, \cdots, k\}$ be the mapping such that for every $w \in V(H)$, $\alpha(w) = \theta(w)$ (i.e. we maintain the same colour for w in H as it was in G). Consider two adjacent vertices u, v in H. These two vertices will also be adjacent in G and thus $\theta(u) \neq \theta(v)$ which implies that $\alpha(u) \neq \alpha(v)$. Thus, α is a k-colouring of H. By definition, $\chi(H) \leq k = \chi(G)$. $\square$

Problem 4. *Construct two graphs H and G such that H is a proper subgraph of G but $\chi(H) = \chi(G)$.*

Solution. One possible solution where $\chi(H) = \chi(G) = 2$ is shown below.

$\square$

Problem 5. *Construct two connected graphs H and G such that H is a spanning subgraph of G and $\chi(H) = \chi(G) - 1$.*

Solution. One possible solution where $\chi(H) = 2 = \chi(G) - 1$ is shown below.

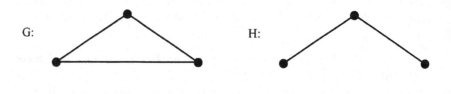

$\square$

Problem 6. *For each of the following graphs, find its chromatic number.*

(a)

(b)

(c)

(d)

Solution.

(a) Let G be the graph. As $e(G) \geq 1, \chi(G) \geq 2$. The following 2-colouring of G shows that $\chi(G) = 2$.

(b) Let G be the graph. As $e(G) \geq 1, \chi(G) \geq 2$. The following 2-colouring of G shows that $\chi(G) = 2$.

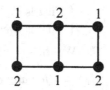

(c) Let G be the graph. As G contains a K_3, $\chi(G) \geq 3$. The following 3-colouring of G shows that $\chi(G) = 3$.

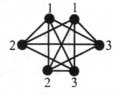

(d) Since every two vertices are adjacent, every vertex in G must be coloured by a new colour. Thus $\chi(G) \geq 5$. The following 5-colouring of G shows that $\chi(G) = 5$.

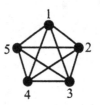

□

Exercise 4.3

Problem 1. *Prove that $\chi(C_n) = 2$ for any even $n \geq 4$.*

Solution. As C_n contains an edge, $\chi(C_n) \geq 2$. Let $n = 2k$ and C_n be $v_1 v_2 \cdots v_{2k} v_1$. Define $\theta : V(C_n) \to \{1, 2\}$ by

$$\theta(v_i) = \begin{cases} 1 \text{ if } i \text{ is odd}; \\ 2 \text{ if } i \text{ is even}. \end{cases}$$

Clearly, θ is a 2-colouring of C_n. We thus conclude that $\chi(C_n) = 2$. □

Problem 2. *Prove that $\chi(C_n) = 3$ for any odd $n \geq 3$.*

Solution. Clearly, $\chi(C_n) \geq 2$. Suppose $\chi(C_n) = 2$ and θ is a 2-colouring of C_n. As before, let C_n be $v_1 v_2 \cdots v_{2k+1} v_1$, where $n = 2k + 1$. We may assume $\theta(v_1) = 1$. Then $\theta(v_2) = 2, \theta(v_3) = 1, \cdots, \theta(v_{2k+1}) = 1$. Observe that $\theta(v_1) = \theta(v_{2k+1}) = 1$ but v_1 and v_{2k+1} are adjacent, which is a contradiction. Thus, $\chi(C_n) \geq 3$. Define $\theta : V(C_n) \to \{1, 2, 3\}$ by

$$\theta(v_i) = \begin{cases} 1 \text{ if } i = n; \\ 2 \text{ if } i \text{ is even}; \\ 3 \text{ otherwise}. \end{cases}$$

Clearly, θ is a 3-colouring of C_n. We thus conclude that $\chi(C_n) = 3$. □

Problem 3. *Find a 3-colouring of the Petersen graph. What is its chromatic number?*

Solution. Let G be the Petersen graph. As G contains a C_5, by the result of Problem 3 in Exercise 4.2, $\chi(G) \geq \chi(C_5) = 3$. The following 3-colouring of G shows that $\chi(G) = 3$.

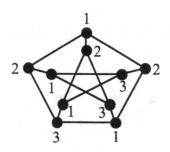

$\square$

Problem 4. *Let G be the graph given below. Explain why $\chi(G) \geq 3$. Then provide a 3-colouring for G, thereby proving that $\chi(G) = 3$.*

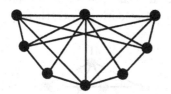

Solution. G contains a K_3 as a subgraph and so $\chi(G) \geq 3$. The following shows a 3-colouring for G, thereby proving that $\chi(G) = 3$.

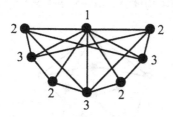

$\square$

Problem 5.   *Let G be the graph given below. Explain why $\chi(G) \geq 4$. Then provide a 4-colouring for G, thereby proving that $\chi(G) = 4$.*

Solution.   Let us label the vertices of G as in the figure below.

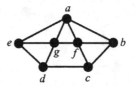

Since G contains a K_3, $\chi(G) \geq 3$. Suppose that $\chi(G) = 3$. We may assume that a, f and g are coloured 1, 2 and 3 respectively. Then e must be coloured 2 and b must be coloured 3. Next, c and d must both be coloured 1. However, c and d are adjacent, which is a contradiction. Thus, $\chi(G) \geq 4$.

The following shows a 4-colouring for G, thereby proving that $\chi(G) = 4$.

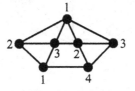

☐

Problem 6. (+)   *Let G be the graph given below. Explain why $\chi(G) \geq 4$. Then provide a 4-colouring for G, thereby proving that $\chi(G) = 4$.*

Solution. Let us label the vertices of the graph G as in the figure below.

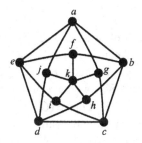

Since G contains an odd cycle (e.g. *abcdea*), $\chi(G) \geq 3$. Suppose $\chi(G) = 3$. By symmetry, we may assume that the vertices a, b, c, d, e of the 5-cycle *abcdea* are coloured 1, 2, 1, 2, 3 in that order as shown below.

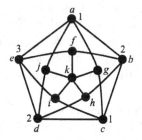

Then f, i and j must be coloured 1, 2 and 3 respectively. However, then k cannot be coloured with any of the 3 colours. Thus, $\chi(G) \geq 4$.

The following shows a 4-colouring for G, thereby proving that $\chi(G) = 4$.

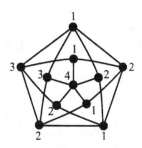

□

Problem 7. (+)   *Determine all graphs G of order $n \geq 2$ with $\chi(G) =$* $n - 1$.

Solution.   Let G be a graph of order $n \geq 2$. We claim that $\chi(G) = n - 1$ if and only if $G \not\cong K_n$ and G contains a K_{n-1} as a subgraph.

Suppose $G \not\cong K_n$ and G contains a K_{n-1}. Then from the result that $\chi(G) = n$ if and only if $G \cong K_n$, we have $\chi(G) < n$. Also, G contains a K_{n-1} implies that $\chi(G) \geq n - 1$. Thus $\chi(G) = n - 1$.

Suppose $\chi(G) = n - 1$. From the result that $\chi(G) = n$ if and only if $G \cong K_n$, we have that $G \not\cong K_n$. We shall show that K_{n-1} is a subgraph of G.

Suppose on the contrary that K_{n-1} is not a subgraph of G. Then $G - x \not\cong K_{n-1}$ for every $x \in V(G)$.

Since $G \not\cong K_n$, there exist $u, v \in V(G)$ such that $uv \notin E(G)$. Then $xy \in E(G)$ for any $x, y \in V(G) \setminus \{u, v\}$; otherwise, $[\{u, v, x, y\}]$ is 2-colourable, implying that $\chi(G) \leq n - 2$, a contradiction. Hence $G - \{u, v\} \cong K_{n-2}$.

Note that $G - u \not\cong K_{n-1}$ and $G - v \not\cong K_{n-1}$. Since $G - \{u, v\} \cong K_{n-2}$, there exist $u', v' \in V(G) \setminus \{u, v\}$ such that $uu', vv' \notin E(G)$.

If $u' = v'$, then u, v and u' can be assigned the same colour, implying that $\chi(G) \leq n - 2$, a contradiction.

If $u' \neq v'$, then the four vertices u, u', v, v' can be coloured by two colours, implying that $\chi(G) \leq n - 2$, a contradiction too.

Therefore K_{n-1} is a subgraph of G.   □

Note. The following is another proof that *if G is of order n and $\chi(G) =$ $n - 1$, then G contains K_{n-1} as a subgraph.*

Let $V(G) = \{v_1, v_2, \ldots, v_n\}$ and θ be a $(n-1)$-colouring of G. We may assume that $\theta(v_i) = i$ for $i \leq n - 1$, $\theta(v_n) = 1$ and $d(v_n) \leq d(v_1)$. Consider the subgraph H induced by $V(G) \setminus \{v_n\}$. Suppose $H \not\cong K_{n-1}$. We shall show that G can be recoloured with a $(n - 2)$-colouring. We have that $v_i v_j \notin E(H)$ for some $i < j \leq n - 1$. If $i \neq 1$, we may recolour v_i as j thus obtaining a $(n-2)$-colouring of G (with colour i excluded), a contradiction. If $i = 1$, we may recolour v_1 as j. Note that $d(v_n) \leq d(v_1) \leq n - 3$. Thus, v_n can be recoloured with at least one colour from $2, 3, \ldots, n - 1$. Thus, G admits a $(n-2)$-colouring (with colour 1 excluded), which is a contradiction. Hence, $H \cong K_{n-1}$ and so G contains a K_{n-1} as a subgraph.   □

Problem 8. *Let G be a graph. Determine whether each of the following statements is true.*

(i) *If G admits a 3-colouring, then G is 3-colourable.*

(ii) *If G is 3-colourable, then G is 5-colourable.*

(iii) *If G is 3-colourable, then $\chi(G) \geq 3$.*

(iv) *If G is 3-colourable, then $\chi(G) \leq 3$.*

(v) *If G is 3-colourable, then G contains an odd cycle.*

(vi) *If G contains an odd cycle, then G is 3-colourable.*

(vii) *If G admits no 3-colourings, then $\chi(G) \geq 3$.*

(viii) *If G admits no 3-colourings, then $\chi(G) = 2$.*

(ix) *If G admits no 3-colourings, then $\chi(G) \leq 2$.*

(x) *If $\chi(G) = 3$, then G contains a triangle.*

(xi) *If $\chi(G) = 3$, then G contains an odd cycle.*

(xii) *If G is a tree with at least two vertices, then $\chi(G) = 2$.*

(xiii) *If $\chi(G) \geq r$, then G contains a K_r as a subgraph.*

Solution.   (i) True.

(ii) True.

(iii) False.

(iv) True

(v) False

(vi) False.

(vii) True. (Indeed, $\chi(G) \geq 4$.)

(viii) False.

(ix) False.

(x) False.

(xi) True.

(xii) True.

(xiii) False.   □

Problem 9.   *Let G be a disconnected graph with two components G_1 and G_2. Show that*

$$\chi(G) = \max\{\chi(G_1), \chi(G_2)\}.$$

Solution.   Let $m = \max\{\chi(G_1), \chi(G_2)\}$. Clearly, $\chi(G) \geq m$.

Let $\alpha : V(G_1) \rightarrow \{1, 2, \cdots, m\}$ be an m-colouring of G_1 and let $\beta : V(G_2) \rightarrow \{1, 2, \cdots, m\}$ be an m-colouring of G_2. Define $\theta : V(G) \rightarrow \{1, 2, \cdots, m\}$ by

$$\theta(v) = \begin{cases} \alpha(v) & \text{if } v \in V(G_1); \\ \beta(v) & \text{if } v \in V(G_2). \end{cases}$$

As there is no edge joining G_1 and G_2, θ is an m-colouring of G; and so $\chi(G) \leq m$. We thus conclude that $\chi(G) = m$. □

Problem 10.   *Let G_1 and G_2 be two connected graphs and let G be the graph obtained from G_1 and G_2 by identifying a vertex in G_1 with a vertex in G_2 as shown below:*

G:

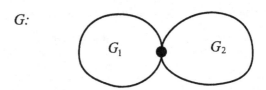

Show that

$$\chi(G) = \max\{\chi(G_1), \chi(G_2)\}.$$

Solution.   Let $m = \max\{\chi(G_1), \chi(G_2)\}$. Clearly, $\chi(G) \geq m$.

Let $x \in V(G_1)$ and $y \in V(G_2)$ be identified in G. Let $\alpha : V(G_1) \rightarrow \{1, 2, \cdots, m\}$ be an m-colouring of G_1 and let $\beta : V(G_2) \rightarrow \{1, 2, \cdots, m\}$ be an m-colouring of G_2. We may assume that $\alpha(x) = \beta(y)$ (why?). Let $\theta : V(G) \rightarrow \{1, 2, \cdots, m\}$ be the mapping defined by

$$\theta(v) = \begin{cases} \alpha(v) & \text{if } v \in V(G_1); \\ \beta(v) & \text{if } v \in V(G_2). \end{cases}$$

Since vertices in $V(G_1) \setminus \{x\}$ are not adjacent to any vertex in $V(G_2) \setminus \{y\}$, θ is an m-colouring of G. Thus, $\chi(G) \leq m$, and so $\chi(G) = m = \max\{\chi(G_1), \chi(G_2)\}$. □

Problem 11.   *Let G be a graph. Show that*

(i) $\chi(G) - 1 \leq \chi(G - v) \leq \chi(G)$ *for each vertex v in G.*

(ii) $\chi(G) - 1 \leq \chi(G - e) \leq \chi(G)$ *for each edge e in G.*

Solution.   (i) Since $G - v$ is a subgraph of G, by the result of Problem 3 in Exercise 4.2, $\chi(G - v) \leq \chi(G)$.

To show that $\chi(G) - 1 \leq \chi(G - v)$, we show that $\chi(G) \leq \chi(G - v) + 1$. Let $k = \chi(G - v)$ and $\theta : V(G - v) \rightarrow \{1, 2, \cdots, k\}$ be a k-colouring of $G - v$. Define $\theta' : V(G) \rightarrow \{1, 2, \cdots, k, k+1\}$ by

$$\theta'(x) = \begin{cases} \theta(x) & \text{if } x \in V(G - v); \\ k+1 & \text{if } x = v. \end{cases}$$

It is clear that θ' is a $(k+1)$-colouring of G. By definition, $\chi(G) \leq k+1 = \chi(G - v) + 1$, as asserted.

(ii) Since $G - e$ is a subgraph of G, by the result of Problem 3 in Exercise 4.2, $\chi(G - e) \leq \chi(G)$. To show that $\chi(G) - 1 \leq \chi(G - e)$, we show that $\chi(G) \leq \chi(G - e) + 1$. Let θ be a k-colouring of $G - e$, where $e = uv$. Define $\theta' : V(G) \rightarrow \{1, 2, \cdots, k, k+1\}$ by

$$\theta'(x) = \begin{cases} \theta(x) & \text{if } x \neq v; \\ k+1 & \text{if } x = v. \end{cases}$$

It is clear that θ' is a $(k+1)$-colouring of G. By definition, $\chi(G) \leq k+1 = \chi(G - e) + 1$, as asserted.   □

Problem 12.   *For each integer $n \geq 2$, construct a graph G of order n such that $\chi(G - v) = \chi(G) - 1$ for each vertex v in G.*

Solution.   The family of complete graphs K_n, $n \geq 2$, satisfies the condition.
□

Problem 13.   *For each integer $n \geq 4$, construct a graph G of order n such that $\chi(G - v) = \chi(G)$ for each vertex v in G.*

Solution.   The family of paths P_n, $n \geq 4$, satisfies the condition.   □

Problem 14.   *For each integer $n \geq 2$, construct a graph G of order n such that $\chi(G - e) = \chi(G) - 1$ for each edge e in G.*

Solution.   The family of complete graphs K_n, $n \geq 2$, satisfies the condition.
□

Problem 15.   *For each integer $n \geq 3$, construct a graph G of order n such that $\chi(G - e) = \chi(G)$ for each edge e in G.*

Solution.   The family of paths P_n, $n \geq 3$, satisfies the condition.   □

Problem 16.   *Let G be the graph shown below:*

(i) *Find $\chi(G)$.*

(ii) *Verify that $\chi(G - e) = \chi(G) - 1$ for each edge e in G.*

Solution.   (i) Let us label the vertices of the graph G as in the figure below.

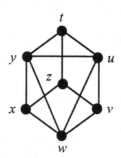

Since G contains a triangle, $\chi(G) \geq 3$. Suppose that $\chi(G) = 3$. We may assume that the vertices u, y, w of the triangle $uywu$ are coloured 1, 2, 3 in that order. Then x, v and t must be coloured 1, 2 and 3, respectively.

However, then z cannot be coloured with any of the 3 colours. Thus, $\chi(G) \geq 4$.

The following shows a 4-colouring for G, thereby proving that $\chi(G) = 4$.

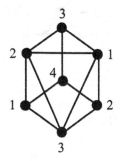

(ii) It suffices to show a 3-colouring of $G - e$ for each edge e in G. We shall do this by stating each edge e to be deleted followed by a listing of the colours of the vertices in the order $t, u, \cdots, z$. By symmetry, we need only consider deleting the edges ty, uy and tz.

$$e = ty \ (yx, xw, wv, vu, ut \text{ are similar}); \ (1, 2, 1, 3, 2, 1, 3)$$
$$e = uy \ (yw, wu \text{ are similar}); \ (1, 2, 1, 3, 1, 2, 2)$$
$$e = tz \ (xz, vz \text{ are similar}); \ (1, 2, 3, 1, 2, 3, 1).$$

$\square$

Problem 17. *Construct a graph G such that $\chi(G) = 3$ and G contains no triangles.*

Solution. G can be the Petersen graph (see Problem 3) or any odd cycle C_n, where $n \geq 5$. $\square$

Problem 18. *Construct a graph G such that $\chi(G) = 4$ and G contains no triangles.*

Solution. The graph of Problem 6 satisfies the condition. $\square$

Problem 19. *Let H be the graph given below. What is the value of* $\chi(H)$?

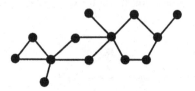

Solution. Since H contains an odd cycle, $\chi(H) \geq 3$. The following 3-colouring of H shows that $\chi(H) = 3$.

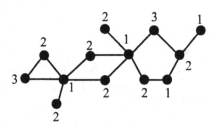

Problem 20. (+)   *Let G be a graph which contains only one odd cycle as a subgraph. Find the value of* $\chi(G)$. *Justify your answer.*

Solution. Since G contains an odd cycle, $\chi(G) \geq 3$.

Let w be any vertex in the only odd cycle of G. Then $G - w$ contains no odd cycle, and so it is a bipartite graph, implying that $\chi(G - w) \leq 2$. By the result of Problem 11 (i), we have $\chi(G) \leq \chi(G - w) + 1 = 3$.

Hence $\chi(G) = 3$.   □

Problem 21. (+)   *Let G be a graph which is not bipartite. Assume that there is a vertex in G which is contained in every odd cycle in G. Show that* $\chi(G) = 3$.

Solution. Since G is not bipartite, $\chi(G) \geq 3$. Let v be a vertex in G which is contained in every odd cycle in G. Then $G - v$ does not contain an odd cycle. Therefore $G - v$ is bipartite and $\chi(G - v) \leq 2$. By the result of Problem 11 (i), we have $\chi(G) \leq \chi(G - v) + 1 = 3$. Thus, $\chi(G) = 3$.   □

Problem 22. (+)   *Let G be a graph. It is known that if $\chi(G) = 3$, then G contains an odd cycle. Assume now that $\chi(G) = 6$. Does G contain two odd cycles which have no vertex in common? Why?*

Solution.   Yes, G must contain two odd cycles which have no vertex in common. We shall prove the contrapositive statement. Suppose G does not contain two odd cycles which have no vertex in common. We shall show that $\chi(G) \leq 5$.

Let C be an odd cycle in G. Then $G - V(C)$ contains no odd cycle. So we can colour $G - V(C)$ with two colours. Next, we colour C with 3 new colours. The resulting colouring of G is a 5-colouring and so, $\chi(G) \leq 5$. Thus, if $\chi(G) = 6$, then G contains two odd cycles which have no vertex in common. $\square$

Problem 23.   *Let G be a graph of order 8 with $\chi(G) = 2$. Show that $e(G) \leq 16$. Construct one such G with $e(G) = 16$.*

Solution.   Since $\chi(G) = 2$, G is bipartite and so the vertices of G can be divided into 2 partite sets, say X and Y. Let $|X| = p$ and so $|Y| = 8 - p$. Since there are no edges between vertices in the same partite set, $e(G) \leq p(8 - p)$. It can be easily verified that the maximum value of $p(8 - p)$ is 16 ($= 4 \times 4$). Thus, $e(G) \leq 16$.

The complete bipartite graph $K(4, 4)$ is one such graph G with $e(G) = 16$. $\square$

Problem 24.   *Let G be a graph of order 7 with $\chi(G) = 3$. Show that $e(G) \leq 16$. Construct one such G with $e(G) = 16$.*

Solution.   Since $\chi(G) = 3$, let X, Y and Z be the sets of vertices in G coloured by colours 1, 2 and 3 respectively. We may assume that $|X| \geq |Y| \geq |Z|$ and note that $(|X|, |Y|, |Z|)$ is one of the following:

$$(5, 1, 1), (4, 2, 1), (3, 3, 1), (3, 2, 2).$$

As there are no edges joining vertices in the same set X (respectively Y, Z), for

$$(5, 1, 1), \ e(G) \leq 11;$$
$$(4, 2, 1), \ e(G) \leq 14;$$
$$(3, 3, 1), \ e(G) \leq 15;$$
$$(3, 2, 2), \ e(G) \leq 16.$$

Thus, $e(G) \leq 16$.

The graph below is one such graph G with $e(G) = 16$.

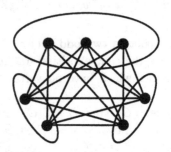

☐

Problem 25.   *Let G be a graph of order 6 with $\chi(G) = 4$. Show that $e(G) \leq 13$. Construct one such G with $e(G) = 13$.*

Solution. Since $\chi(G) = 4$, let W, X, Y and Z be the sets of vertices in G coloured by colours 1, 2, 3 and 4 respectively. We may assume that $|W| \geq |X| \geq |Y| \geq |Z|$ and note that $(|W|, |X|, |Y|, |Z|)$ is either $(3, 1, 1, 1)$ or $(2, 2, 1, 1)$. As there are no edges joining vertices in the same set W (respectively X, Y, Z), for

$$(3, 1, 1, 1), \quad e(G) \leq 12; \quad (2, 2, 1, 1), \quad e(G) \leq 13.$$

Thus, $e(G) \leq 13$.

The graph below is one such graph G with $e(G) = 13$.

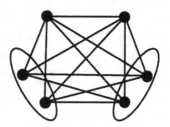

☐

Problem 26. *Determine the chromatic number of each of the following graphs:*

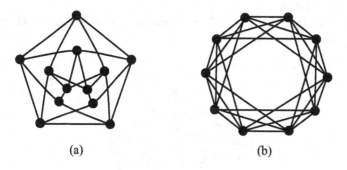

(a)   (b)

Solution. Let us label the vertices of the graph G as in the figure below.

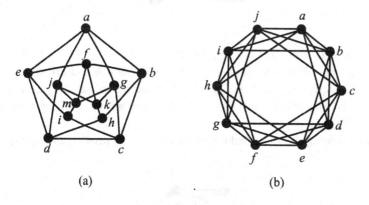

(a)   (b)

(a) Since G contains an odd cycle (e.g. *abcdea*), $\chi(G) \geq 3$. Suppose $\chi(G) = 3$. Then G is 3-colourable. Let $\theta : V(G) \to \{1,2,3\}$ be a 3-colouring of G. Consider the 5-cycle *abcdea*. By symmetry, we may assume that $(\theta(a), \theta(b), \theta(c), \theta(d), \theta(e))$ is one of the following: $(1,2,1,2,3), (1,2,1,3,2), (1,2,3,2,3)$.

If $(\theta(a), \theta(b), \theta(c), \theta(d), \theta(e)) = (1,2,1,2,3)$, then f, i and j must be coloured 1, 2 and 3 respectively. However, now m cannot be coloured with any of the 3 colours.

If $(\theta(a), \theta(b), \theta(c), \theta(d), \theta(e)) = (1,2,1,3,2)$, then h, j and i must be coloured 1, 2 and 3 respectively. Now, in turn, m must be coloured 1 and f must be coloured 3. However, now k cannot be coloured with any of the 3 colours.

If $(\theta(a), \theta(b), \theta(c), \theta(d), \theta(e)) = (1, 2, 3, 2, 3)$, then f, g and j must be coloured 1, 2 and 3 respectively. However, now m cannot be coloured with any of the 3 colours.

Thus, $\chi(G) \geq 4$. The following shows a 4-colouring for G, thereby proving that $\chi(G) = 4$.

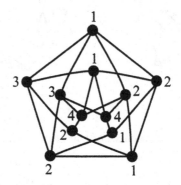

(b) Since G contains a K_4, $\chi(G) \geq 4$. Suppose $\chi(G) = 4$. Then G is 4-colourable. We may assume that the vertices a, b, c, d are coloured 1, 2, 3, 4 respectively. Then e, f, g, and h must be coloured 1, 2, 3 and 4 respectively. However, then i cannot be coloured with any of the 4 colours. Thus, $\chi(G) \geq 5$.

The following shows a 5-colouring for G, thereby proving that $\chi(G) = 5$.

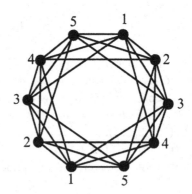

□

Problem 27.   *Let G and H be two graphs. The* **join** *of G and H, denoted by $G + H$, is the graph whose vertex set is the union of $V(G)$ and $V(H)$, and whose edge set consists of the edges in G and H together with new edges which join every vertex in G to every vertex in H. Thus, if A and B are the graphs shown below:*

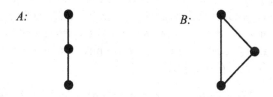

A: *B:*

then $A + B$ is the graph shown below:

A+B:

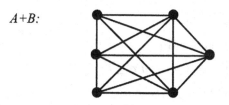

(i) Find $\chi(A), \chi(B)$ and $\chi(A + B)$.

(ii) In general, what is the relation among $\chi(G), \chi(H)$ and $\chi(G + H)$? Prove your result.

Solution.   (i) It is clear that $\chi(A) = 2$ and $\chi(B) = 3$.

Since $A + B$ contains a K_5, $\chi(A + B) \geq 5$. The following shows a 5-colouring for $A + B$, thereby proving that $\chi(A + B) = 5$.

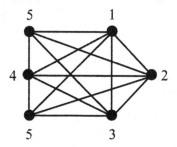

(ii) In general, $\chi(G+H) = \chi(G)+\chi(H)$. We prove the result as follows. The subgraph G in $G+H$ needs at least $\chi(G)$ colours. Since the vertices of the subgraph H are each adjacent to all the vertices of G in $G+H$, a valid colouring of $G+H$ would require colours different from those used for G to colour the vertices in H. At least $\chi(H)$ new colours are needed for H in $G+H$. Thus, $\chi(G+H) \geq \chi(G)+\chi(H)$. In fact, any valid $\chi(G)$-colouring of G with $\chi(G)$ colours followed by any valid $\chi(H)$-colouring of G with $\chi(H)$ colours different from those used for the vertices of G will result in a valid $(\chi(G)+\chi(H))$-colouring of $G+H$. Thus, $\chi(G+H) \leq \chi(G)+\chi(H)$, and so $\chi(G+H) = \chi(G)+\chi(H)$. $\square$

Problem 28.   *The wheel of order n, denoted by W_n, is defined as (see Problem 27 above) $W_n = C_{n-1} + K_1$.*

(i) *Draw W_6 and W_7.*

(ii) *Find a 3-colouring for W_7.*

(iii) *Find a 4-colouring for W_6.*

(iv) *Show that $\chi(W_n) = 3$ for odd $n \geq 5$.*

(v) *Show that $\chi(W_n) = 4$ for even $n \geq 4$.*

Solution.   (i)

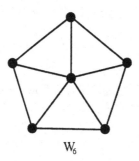 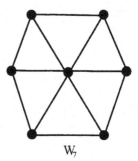

W_6 W_7

(ii)

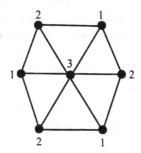

(iii)

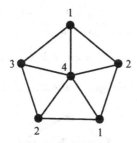

(iv) For odd $n \geq 5$, $n-1$ is even. Thus, since $W_n = C_{n-1} + K_1$, we have

$$\chi(W_n) = \chi(C_{n-1}) + \chi(K_1) = 2 + 1 = 3.$$

(v) For even $n \geq 4$, $n-1$ is odd. Thus, since $W_n = C_{n-1} + K_1$, we have

$$\chi(W_n) = \chi(C_{n-1}) + \chi(K_1) = 3 + 1 = 4.$$

$\square$

Problem 29. (+)   *Let G be a graph of order $n \geq 5$ which contains a P_5 as an induced subgraph. Show that $\chi(G) \leq n-3$. For each $n \geq 5$, construct one such G of order n for which the equality $\chi(G) = n-3$ holds.*

Solution.   Let $G' = G - V(P_5)$. As $v(G') = n-5$, G' can be coloured with at most $n-5$ colours. Since P_5 is an induced subgraph of G, we may colour P_5 by two new colours. Hence $\chi(G) \leq (n-5) + 2 = n-3$.

The graph $K_{n-5} + P_5$ is a graph of order n which contains a P_5 as an induced subgraph and $\chi(K_{n-5} + P_5) = (n-5) + 2 = n-3$. $\square$

Problem 30.   *Let G be a graph. A set of vertices S in G is said to be* **independent** *if no two vertices in S are adjacent.*

Assume that $\chi(G) = k$ and there is a k-colouring θ of G. For each $i = 1, 2, \cdots, k$, let V_i be the set of vertices v in G with $\theta(v) = i$.

(i) *Can V_i be empty?*

(ii) *Is V_i an independent set?*

Solution. (i) No, V_i cannot be empty for any $i = 1, 2, \cdots, k$; otherwise, $\chi(G) < k$, a contradiction.

(ii) Yes, V_i is an independent set; otherwise, two adjacent vertices in V_i will have to be coloured differently.   □

Problem 31. (∗)   *Let G be a graph of order n. The* **independence number** *of G, denoted by $\alpha(G)$, is defined by*

$$\alpha(G) = \max\{|S| \ \mid \ S \text{ is an independent set in } G\}.$$

(i) *Find $\alpha(H)$, where H is the graph shown below:*

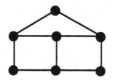

(ii) *Show that $\chi(G)\alpha(G) \geq n$.*

(iii) *Construct a connected graph H such that $v(H) = 12$, $\chi(H) = 4$ and $\alpha(H) = 3$.*

(iv) *Show that $\chi(G) + \alpha(G) \leq n + 1$.*

(v) *Construct a connected graph H such that $v(H) = 11$, $\chi(H) = 5$ and $\alpha(H) = 7$.*

Solution. For the solutions below, unless otherwise stated, let V_i be the set of vertices v in G with $\theta(v) = i$ for each $i = 1, 2, \cdots, \chi(G)$, where θ is a $\chi(G)$-colouring of the graph G.

(i) The set of 4 'white' vertices indicated in the figure below is an independent set in H. It is easy to check that there is no independent set in H with 5 vertices. Thus, $\alpha(H) = 4$.

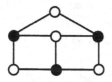

(ii) Since $\alpha(G) \geq \max\{|V_i| \mid i = 1, 2, \cdots, \chi(G)\}$, we have

$$\chi(G)\alpha(G) \geq \sum_{i=1}^{\chi(G)} |V_i|,$$

i.e., $\chi(G)\alpha(G) \geq n$.

(iii) The following connected graph H is such that $v(H) = 12$, $\chi(H) = 4$ and $\alpha(H) = 3$.

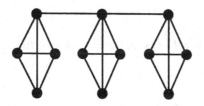

(iv) Let A be an independent set in G such that $|A| = \alpha(G)$. Then the number of vertices in $G - A$ is $n - \alpha(G)$. Introduce a colouring of G as follows: all vertices in A are coloured by one colour, and each of the $n - \alpha(G)$ vertices in $G - A$ is coloured by one new colour. Clearly, this defines a $(n + 1 - \alpha(G))$-colouring of G. Thus, $\chi(G) \leq n + 1 - \alpha(G)$ and so $\chi(G) + \alpha(G) \leq n + 1$.

(v) $K_4 + N_7$ (see below) is a connected graph with the required properties.

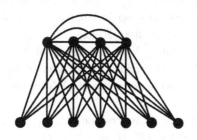

□

Problem 32. (∗)   *Let G be a graph which is not bipartite. Assume that G contains an independent set S such that $V(C) \cap S$ is non-empty for every odd cycle C in G. Show that $\chi(G) = 3$.*

Solution.   Since G is not bipartite, $\chi(G) \geq 3$. By assumption, $G - S$ does not contain an odd cycle. Therefore $G - S$ is bipartite and is 2-colourable. It is possible to colour all vertices in S with one colour because S is independent. Thus G has a 3-colouring. Hence, $\chi(G) = 3$.   □

Problem 33. (+)   *Let G be a graph satisfying the following conditions:*
(1) $\chi(G) = 5$ and
(2) $\chi(G - v) = 4$ for each vertex v in G.
Show that

(i) *G is connected;*

(ii) *$\delta(G) \geq 4$;*

(iii) *$N(u)$ is not a subset of $N(v)$ for any two vertices u, v in G;*

(iv) *$v(G) \neq 6$.*

Solution.   (i) Let $G = \bigcup_{i=1}^{k} H_i$, where $k \geq 2$ and the H_i's are the components of G. Then $\chi(G) = \max\{\chi(H_i) | i = 1, 2, \cdots, k\}$. We may assume that $\chi(G) = \chi(H_1) = 5$. Then, $\chi(G - v) = \chi(H_1) = 5$ for each vertex v in H_2, a contradiction. Thus, G is connected.

(ii) Suppose $\delta(G) \leq 3$. Let v be a vertex in G such that $d(v) = \delta(G)$. Since $\chi(G - v) = 4$, there is a colouring $\theta : V(G - v) \to \{1, 2, 3, 4\}$. Since $d(v) = \delta(G) \leq 3$, there is at least one colour, say 4, not used to colour the neighbours of v. Extend θ to colour G by colouring v the colour 4. This gives a 4-colouring of G. Thus, $\chi(G) \leq 4$, a contradiction. Hence $\delta(G) \geq 4$.

(iii) Suppose there exist two vertices u, v in G such that $N(u)$ is a subset of $N(v)$ (then u and v are non-adjacent). Since $\chi(G - u) = 4$, there is a 4-colouring θ of $G - u$. Since $N(u)$ is a subset of $N(v)$, θ can be extended to a 4-colouring of G by colouring u the colour $\theta(v)$. Thus $\chi(G) = 4$, a contradiction. Hence $N(u)$ is not a subset of $N(v)$ for any two vertices u, v in G.

(iv) Suppose $v(G) = 6$. From Problem 7, $G \neq K_6$ and G contains a K_5. Then there exists a vertex v in G such that $G - v = K_5$. However, this means that $\chi(G - v) = 5$, a contradiction. Thus, $v(G) \neq 6$.   □

Exercise 4.4

Problem 1. *Let G be the graph C_6 as shown below with two different ways of arranging its vertices. Apply the greedy colouring algorithm to colour G and find the number of colours produced in each case.*

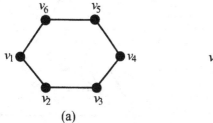

(a)   (b)

Solution. The figure below shows the result of applying the greedy algorithm. The number of colours in case (a) and case (b) is 2 and 3 respectively.

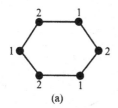 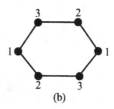

(a)   (b)

□

Problem 2. *Let G be the graph as shown below with two different ways of arranging its vertices. Apply the greedy colouring algorithm to colour G and find the number of colours produced in each case.*

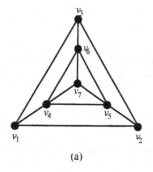 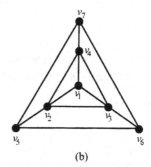

(a)   (b)

Solution. The figure below shows the result of applying the greedy algorithm. The number of colours in both cases is 4.

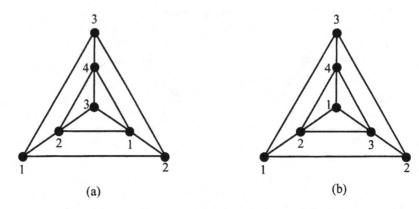

(a)   (b)

Problem 3. *Let G be the graph given below:*

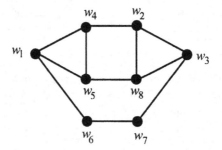

(i) *Find the number of colours produced by applying the greedy colouring algorithm on G according to the ordering of vertices $w_1, w_2, \cdots, w_8$.*

(ii) *Find $\chi(G)$.*

Solution. (i) The figure below shows the result of applying the greedy algorithm. The number of colours produced is 4.

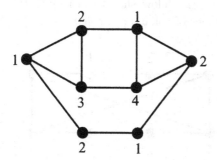

(ii) Since G contains a triangle, $\chi(G) \geq 3$. The figure below shows a 3-colouring of G, proving that $\chi(G) = 3$.

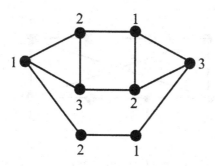

$\square$

Problem 4.   *Let G be the graph given below:*

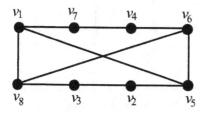

(i) *Find the number of colours produced by applying the greedy colouring algorithm on G according to the ordering of vertices $v_1, v_2, \cdots, v_8$.*

(ii) *Find $\chi(G)$.*

Solution.   (i) The figure below shows the result of applying the greedy algorithm. The number of colours produced is 4.

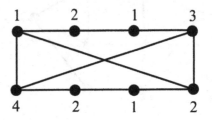

(ii) As G contains a C_5, $\chi(G) \geq 3$. The figure below shows a 3-colouring of G, proving that $\chi(G) = 3$.

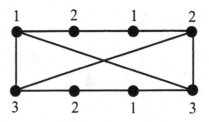

□

Problem 5.   *Let H be the graph given below:*

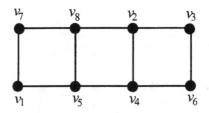

(i) *Find the number of colours produced by applying the greedy colouring algorithm on H according to the ordering: $v_1, v_2, \cdots, v_8$.*

(ii) *Determine $\chi(H)$.*

Solution.   (i) The figure below shows the result of applying the greedy algorithm. The number of colours produced is 4.

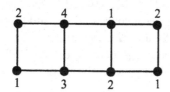

(ii) Since G is a bipartite graph with at least one edge, $\chi(G) = 2$. ☐

Problem 6. *Let H be the graph given below:*

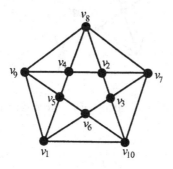

(i) *Find the number of colours produced by applying the greedy colouring algorithm on H according to the ordering:* $v_1, v_2, \cdots, v_{10}$.

(ii) *Determine $\chi(H)$.*

Solution. (i) The figure below shows the result of applying the greedy algorithm. The number of colours produced is 5.

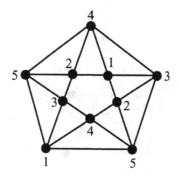

(ii) Since G contains a triangle, $\chi(G) \geq 3$. Suppose $\chi(G) = 3$. We may assume that v_8, v_4 and v_2 are coloured '1', '2' and '3' respectively. Then v_9 and v_7 must be coloured '3' and '2' respectively. Next, both v_5 and v_3 must be coloured '1'. Now v_1 and v_{10} must be coloured '2' and '3' respectively. But then v_6 cannot be coloured with '1', '2' or '3'. Thus, $\chi(G) \geq 4$.

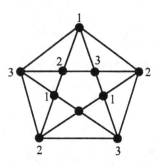

The figure below shows a 4-colouring of G, proving that $\chi(G) = 4$.

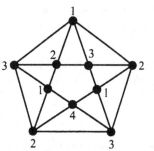

Problem 7. *Let H be the graph given below in which the vertices are named.*

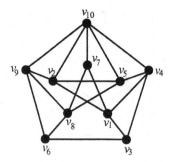

(i) *Find the number of colours produced by applying the greedy colouring algorithm on H according to the ordering: $v_1, v_2, \cdots, v_{10}$.*

(ii) *Determine $\chi(H)$.*

Solution. (i) The figure below shows the result of applying the greedy algorithm. The number of colours produced is 5.

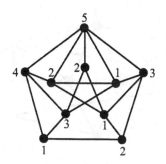

(ii) Since G contains a triangle, $\chi(G) \geq 3$. The figure below shows a 3-colouring of G, proving that $\chi(G) = 3$.

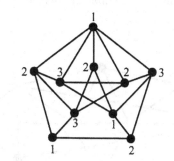

□

Problem 8. (+) *Let G be the graph given below.*

(i) *Find $\chi(G)$.*

(ii) *Arrange the vertices as $v_1, v_2, \cdots, v_{10}$ so that, when the greedy colouring algorithm is applied to G according to this ordering, the number of colours produced is the value of $\chi(G)$.*

Solution.   (i) Since G contains a triangle, $\chi(G) \geq 3$. The figure below shows a 3-colouring of G, proving that $\chi(G) = 3$.

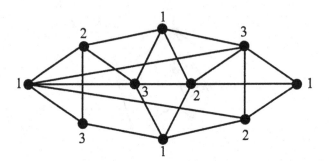

(ii) The figure below shows an arrangement of the vertices as $v_1, v_2, \cdots, v_{10}$ so that, when the greedy colouring algorithm is applied to G according to this ordering, the number of colours produced is the value of $\chi(G)$.

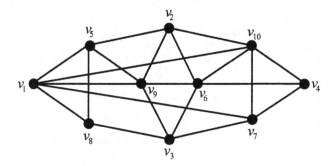

□

Exercise 4.5

Problem 1. *Determine the chromatic number of the following graph. Prove your result.*

Solution. Since the graph G contains K_4, $\chi(G) \geq 4$. Since G is connected and is neither a complete graph nor an odd cycle, by Brooks' Theorem, $\chi(G) \leq \Delta(G) = 4$. Thus, $\chi(G) = 4$. $\qquad\qquad\square$

Problem 2. *Determine the chromatic number of the following graph. Prove your result.*

Solution. Since the graph G contains a triangle, $\chi(G) \geq 3$. The 3-colouring of G below shows that $\chi(G) = 3$.

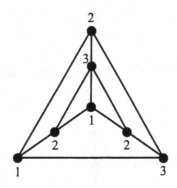

Problem 3. (+)   *Let G be a connected graph, which is not a complete graph. Show that if G contains K_r as a subgraph, where $r = \Delta(G) \geq 3$, then $\chi(G) = r$.*

Solution.   Since the graph G contains K_r, $\chi(G) \geq r$. Since G is connected and is neither a complete graph nor an odd cycle (why?), by Brooks' Theorem, $\chi(G) \leq \Delta(G) = r$. Thus, $\chi(G) = r$.   □

Problem 4. (+)   *Consider the following graph H:*

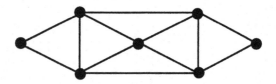

(i) *Find $\Delta(H)$ and $\chi(H)$.*

(ii) *Arrange the vertices of H as $v_1, v_2, \cdots, v_7$ so that, when the greedy algorithm is applied to H according to this ordering, the number of colours produced is $\Delta(H) + 1$.*

Solution.   (i) $\Delta(H) = 4$. Since H contains a triangle, $\chi(H) \geq 3$. The 3-colouring of H below shows that $\chi(H) = 3$.

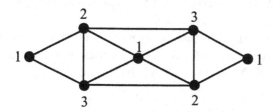

(ii) The figure below shows an arrangement of the vertices of H as $v_1, v_2, \cdots, v_7$ so that, when the greedy algorithm is applied to H according to this ordering, the number of colours produced is $\Delta(H) + 1 = 5$. The colours are indicated in parentheses.

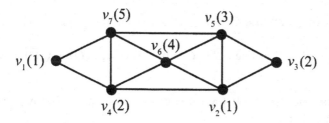

□

Problem 5. (+)   *Does there exist a graph G satisfying the following conditions:*

(i) $\chi(G) = 7$ *and*

(ii) *the degree sequence of G is* $(6, 6, 6, 6, 6, 5, 5, 5, 5, 4, 4, 3, 3, 3, 3)$?

Solution.   No, such a G does not exist. We shall prove it by contradiction.
    Suppose such a G exists. Let H be a component of G with $\chi(H) = \chi(G) = 7$. Clearly, H is not an odd cycle. If H is a complete graph, then by (ii), $H \cong K_r$, where $r \le 6$. But then $\chi(H) = \chi(K_r) = r \le 6$, a contradiction. Thus, H is a connected graph which is neither an odd cycle nor a complete graph. By Brooks' Theorem, $\chi(H) \le \Delta(H) \le 6$, a contradiction again. Thus, no such G exists.   □

Problem 6. (+)   *Let G be a cubic connected graph. What are the possible values of $\chi(G)$? Classify G according to the value of $\chi(G)$.*

Solution. As G is connected, $\chi(G) \geq 2$. By Theorem 4.1, $\chi(G) \leq \Delta(G) + 1 = 3 + 1 = 4$. Thus, $2 \leq \chi(G) \leq 4$.

If $G \cong K_4$ (3-regular), then $\chi(G) = 4$. If $G \not\cong K_4$, then G can never be complete. As G is cubic, G is not an odd cycle. Hence, by Brook's Theorem, $\chi(G) \leq \Delta(G) = 3$. We thus have the following conclusion:

$$\chi(G) = \begin{cases} 4 & \text{if } G \cong K_4; \\ 2 & \text{if } G \text{ is bipartite;} \\ 3 & \text{otherwise.} \end{cases}$$

$\square$

Problem 7. (+) *Let G be a regular and connected graph of order n. Show that $\chi(G) + \chi(\overline{G}) = n + 1$ if and only if $G \cong K_n$ or $G \cong C_5$.*

Solution. ($\Leftarrow$) Suppose $G \cong K_n$. Then $\overline{G} \cong N_n$. Thus, $\chi(G) + \chi(\overline{G}) = n + 1$.

Suppose $G \cong C_5$. Then $\overline{G} \cong C_5$. In this case, $\chi(G) + \chi(\overline{G}) = 3 + 3 = 5 + 1$.

($\Rightarrow$) Assume that $\chi(G) + \chi(\overline{G}) = n + 1$ and suppose $G \not\cong K_n$. We shall prove that $G \cong C_5$.

Let G be k-regular. Then $2 \leq k \leq n - 2$ and $\overline{G}$ is $(n - 1 - k)$-regular. Assume that G is not an odd cycle. Then by Brook's Theorem, $\chi(G) \leq \Delta(G) = k$. As $\chi(\overline{G}) \leq \Delta(\overline{G}) + 1 = (n - 1 - k) + 1 = n - k$ by Theorem 4.1, we have

$$\chi(G) + \chi(\overline{G}) \leq k + (n - k) = n,$$

a contradiction. Thus $G \cong C_{2r+1}$ (here, $n = 2r + 1$). If $r = 1$, then $G \cong K_3$, which is not allowed. Assume $r \geq 3$. Then $\overline{G}$ $(= \overline{C_{2r+1}})$ is a connected graph which is neither complete nor an odd cycle. By Brook's Theorem, $\chi(\overline{G}) \leq \Delta(\overline{G}) = (n - 1) - 2 = n - 3$. Thus,

$$\chi(G) + \chi(\overline{G}) \leq 3 + (n - 3) = n,$$

a contradiction.

Hence $r = 2$, and so $G \cong C_5$, as required.

$\square$

Exercise 4.6

Problem 1. *A chemist wishes to ship chemicals* A, B, C, D, W, X, Y, Z *using as few containers as possible. Certain chemicals cannot be shipped in the same container since they will react with each other. In particular, any two of the chemicals in each of the following 6 groups*

$$\{A, B, C\}, \{A, B, D\}, \{A, B, X\}, \{C, W, Y\}, \{C, Y, Z\} \text{ and } \{D, W, Z\}$$

react with each other. Draw a graph to model these relations between the chemicals. Use this graph to find the minimum number of containers needed to ship the chemicals. Is it possible to have an allocation of the chemicals that uses the minimum number of containers and such that there are at most two chemicals in each container?

Solution. Let G be the graph with 8 vertices which represent the 8 chemicals A, B, C, D, W, X, Y, Z, where two vertices are adjacent if and only if the chemicals they represent react with each other.

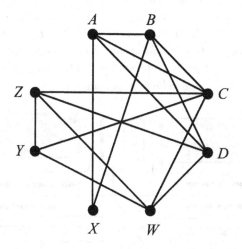

Since the subgraph induced by the vertices C, W, Y and Z is K_4, $\chi(G) \geq$ 4. The figure below shows a 4-colouring of G where each colour is used exactly two times. Thus, the minimum number of containers needed is 4 and a suitable allocation of the chemicals such that there are at most two chemicals in each container is $\{C, X\}, \{B, W\}, \{D, Y\}, \{A, Z\}$.

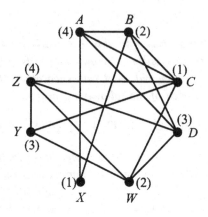

□

Problem 2.   *The following figure shows the intersection of a major road and a small road. There are 10 traffic lanes, L_1 to L_{10}, along which vehicles approach the intersection. The directions in which vehicles along each of the lanes are allowed to negotiate the intersection and go on to a prescribed exit lane are shown. A traffic light system is installed to control movement through the intersection. The system consists of a certain number of phases. At each phase, vehicles in lanes for which the light is green may proceed safely through the intersection. What is the minimum number of phases needed for the traffic light system so that (eventually) all vehicles may proceed safely through the intersection? (We may assume that each lane is broad enough for one vehicle at a time.)*

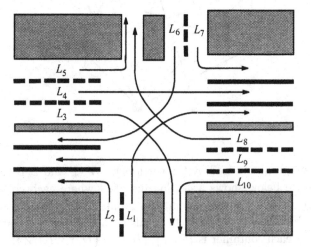

Solution. Let G be the graph with 10 vertices which represent the 10 lanes, where two vertices are adjacent if and only if vehicles travelling in the directions allowed by the two lanes represented by the vertices will collide with each other.

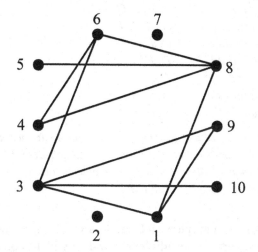

Since the subgraph induced by the vertices 1, 3 and 9 is K_3, $\chi(G) \geq 3$. The figure below shows a 3-colouring θ of G, using the colours red (r), green (g) and yellow (y). Thus, the minimum number of phases needed for the traffic light system is 3 and a possible allocation of the phases, based on the colouring θ, is $\{L_1, L_2, L_6, L_7\}, \{L_4, L_5, L_9, L_{10}\}, \{L_3, L_8\}$.

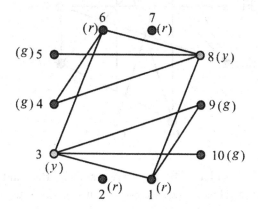

Problem 3.   *A Student Council has 8 committees.   Ten councilors $A, B, C, D, E, F, G, H, I, J$ are appointed to be members of the committees as shown below:*

Publicity :	A, B, C, D
Recreation :	A, E, F, G
Welfare :	G, H, I, J
School Liaison :	C, J
Community :	D, E
Projects :	A, C
Secretariat :	B, F, H
Finance :	G, I

If each committee is scheduled to meet for two hours each week, what is the smallest number of two-hour sessions required to schedule all 8 committee meetings so that each of these councilors is able to attend the meetings of the committees he/she is a member of?

Solution.   Let Z be the graph with 8 vertices which represent the 8 committees, where two vertices are adjacent if and only if the committees they represent have a common member.

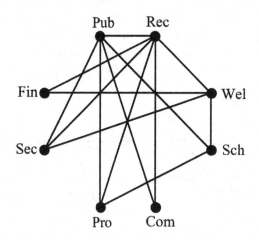

Since the subgraph induced by the vertices Pub, Rec and Com is K_3, $\chi(Z) \geq 3$. The figure below shows a 3-colouring of Z. Thus, the minimum number of two-hour timeslots needed is 3 and a suitable time allocation of timeslots is {Pub, Wel}, {Rec, Sch}, {Com, Pro, Sec, Fin}.   □

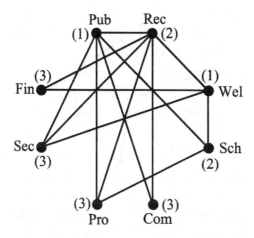

Problem 4. *A school is preparing a timetable for exams in 7 different subjects, labelled A to G. It is understood that if there is a pupil taking two of these subjects, their exams must be held in different timeslots. The table below shows (by crosses) the pairs of subjects which are taken by at least one pupil in common. The school wants to find the minimum number of timeslots necessary and also to allocate subjects to the timeslots accordingly. Interpreting this problem as a vertex-colouring problem, find the minimum number of timeslots needed and a suitable time allocation of the subjects.*

	A	B	C	D	E	F	G
A		X	X	X		X	
B	X		X			X	X
C	X	X		X			X
D	X		X		X		
E				X		X	X
F	X	X			X		X
G		X	X		X	X	

Solution. Let Z be the graph with 7 vertices which represent the 7 subjects A, B, C, D, E, F and G, where two vertices are adjacent if and only if there is a pupil taking the two subjects represented by the vertices.

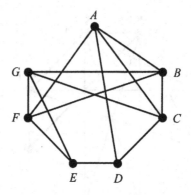

Since Z contains a triangle, $\chi(Z) \geq 3$. Suppose $\chi(Z) = 3$. Let $\theta :$ $V(Z) \to \{1, 2, 3\}$ be a 3-colouring of Z. We may assume that the vertices A and B are coloured '1' and '2' respectively. Then C and F must both be coloured '3'. Next, G and E must be coloured '1' and '2' in turn. However, D now cannot be coloured with '1', '2' or '3', a contradiction. Thus, $\chi(Z) \geq 4$. The figure below shows a 4-colouring of Z. Thus, the minimum number of timeslots needed is 4 and a suitable time allocation of the subjects is $\{A, G\}$, $\{B, E\}$, $\{C, F\}$ and $\{D\}$.

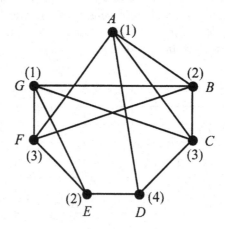

□

Chapter 5

Matchings in Bipartite Graphs

Theorem 5.1 *Let G be a bipartite graph with bipartition (X, Y). Then G contains a complete matching from X to Y if and only if $|S| \leq |N(S)|$ for every subset S of X.* □

Corollary 5.2 *Let G be a bipartite graph with bipartition (X, Y) such that $|X| = |Y|$. Then G has a perfect matching if and only if $|S| \leq |N(S)|$ for every subset S of X.* □

Corollary 5.3 *Every k-regular bipartite graph, where $k \geq 1$, always contains a perfect matching.* □

Theorem 5.4 *The family $(S_1, S_2, \cdots, S_m)$ of non-empty finite subsets of a set W has an SDR if and only if*

$$\left| \bigcup_{i \in I} S_i \right| \geq |I|$$

for all subsets I of $\{1, 2, \cdots, m\}$. □

Exercise 5.2

Problem 1. *Five applicants $A_1, A_2, \cdots, A_5$ apply for five jobs $J_1, J_2,$ $\cdots, J_5$. It is known that*

(i) *J_1 is applied only by A_2,*

(ii) *J_2 is applied by all except A_4,*

(iii) *J_3 is applied by all except A_2,*

(iv) *J_4 is applied by A_2 and A_4, and*

(v) *J_5 is applied only by A_4.*

(a) Draw a bipartite graph that models the situation.

(b) Is it possible to assign each applicant to a job for which he/she applies?

Solution. (a) The bipartite graph is shown below:

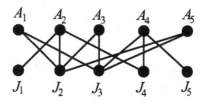

(b) It is impossible. If it is possible to assign each applicant to a job for which he/she applies, then all the five jobs must be assigned to the five applicants. Thus J_1 must be assigned to A_2 and J_5 must be assigned to A_4. But, then J_4 cannot be assigned to any one, a contradiction. $\square$

Problem 2. *In the preceding problem, suppose that the applicant A_5 changes his/her mind and applies for J_5 instead of J_2.*

(a) Draw a bipartite graph that models the situation.

(b) Is it possible to assign each applicant to a job for which he/she applies?

Solution. (a) The bipartite graph is shown below:

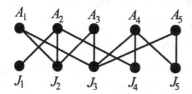

(b) It is possible, and the arrangment is shown below:

$$J_1 - A_2; J_2 - A_1; J_3 - A_3; J_4 - A_4; J_5 - A_5.$$

$\square$

Problem 3. *Five applicants apply to work in a company. There are six jobs available: $J_1, J_2, \cdots, J_6$. Applicant A is qualified for jobs J_2 and J_6; B is qualified for jobs J_1, J_3 and J_4; C is qualified for jobs J_2, J_3 and J_6; D is qualified for jobs J_1, J_2 and J_3; E is qualified for all jobs except J_4 and J_6.*

(a) Draw a bipartite graph that models the situation.

(b) Is it possible to assign each applicant to a job for which he/she is qualified?

Solution.   (a) The bipartite graph is shown below:

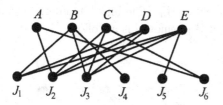

(b) It is possible, and the arrangment is shown below:

$$J_6 - A; J_1 - B; J_2 - C; J_3 - D; J_5 - E.$$

$\square$

Problem 4. *Five men $M_1, M_2, \cdots, M_5$ and five women $W_1, W_2, \cdots, W_5$ have and only have the following acquaintance relationships between them:*

(i) *each of W_1, W_2 and W_3 is acquainted with all the men,*

(ii) *each of M_1 and M_5 is acquainted with all the women.*

(a) Draw a bipartite graph that models the situation.

(b) Is it possible to marry off these five men in such a way that each man marries a woman he is acquainted with?

(c) If M_1 insists on marrying W_1, is it possible to marry off the remaining ones in such a way that each man marries a woman he is acquainted with?

Solution. (a) The bipartite graph that models the situation is shown below:

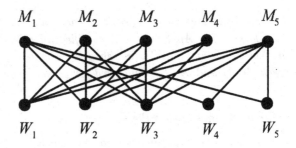

(b) Yes. The following matching will work:

$$\{M_1W_4, M_2W_2, M_3W_3, M_4W_1, M_5W_5\}.$$

(c) Suppose M_1 marries W_1. The bipartite graph that models the new situation with M_1 and W_1 removed is as follows:

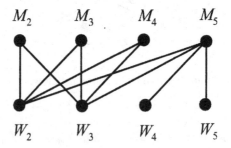

Observe that three men M_2, M_3 and M_4 are only acquainted with two women W_2 and W_3. It is thus not possible to marry off the remaining ones in such a way that each man marries a woman he is acquainted with. □

Problem 5. *Consider the following set of codewords:*

$$X = \{ab, abc, cd, bcd, de\}.$$

We wish to transmit these codewords as messages. Instead of transmitting the whole codeword, we transmit a single letter which is contained in it, as its representative. Can this be done in such a way that the five codewords can be recovered uniquely from their five respective representatives?

Solution. We construct a bipartite graph such that the partite sets are the set of codewords $\{ab, abc, cd, bcd, de\}$ and the set of letters $\{a, b, c, d, e\}$, and a codeword is adjacent only to the letters contained in it.

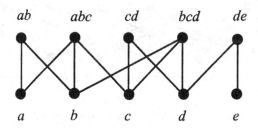

Single letters can be transmitted in such a way that the five codewords can be recovered uniquely from their five respective representatives. The following matching of codewords and representatives will work:

$$\{(ab, b), (abc, a), (cd, d), (bcd, c), (de, e)\}.$$

$\square$

Problem 6. *A school has vacancies for seven teachers, one for each of the subjects Chemistry, English, French, Geography, History, Mathematics and Physics. There are seven applicants for the vacancies and all are qualified to teach more than one subject. The applicants and their subjects are listed in the table below.*

(a) Draw a bipartite graph to represent this situation.

(b) Determine the maximum number of (suitably qualified) teachers the school can employ.

Applicants	Subjects qualified
Miss Lim	Mathematics, Physics
Miss Wong	Chemistry, English, Mathematics
Miss Tay	Chemistry, French, History, Physics
Mr. Tan	English, French, History, Physics
Mr. Lee	Chemistry, Mathematics
Mr. Ng	Mathematics, Physics
Mr. Peng	English, Geography, History

Solution. (a) The bipartite graph that models the situation is shown below:

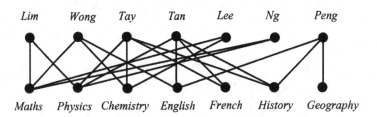

(b) The school can employ all 7 of the applicants. The following job assignment will work:

{(Lim, Maths), (Wong, English), (Tay, French), (Tan, History), (Lee, Chemistry), (Ng, Physics), (Peng, Geography)}.   □

Problem 7. *Consider the following bipartite graph G with bipartition* (X, Y):

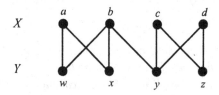

(i) *Is* $\{bx, cy\}$ *a matching?*

(ii) *Is* $\{ax, by, cy\}$ *a matching?*

(iii) *Is* $\{ax, by, cz\}$ *a matching?*

(iv) *Is* $\{ax, bw, cz, dy\}$ *a matching? a perfect matching?*

(v) *Is there any perfect matching that contains the edge 'by'?*

(vi) *Find the number of perfect matchings in G.*

Solution. (i) Yes.

(ii) No, since the edges by and cy are incident with a common vertex y.

(iii) Yes.

(iv) Yes. Yes.

(v) No. To see this, let us assume that there is a perfect matching that contains the edge by. Then the matching must contain the edge cz. However, this means that the vertex d is not contained in any edge of the perfect matching, which is a contradiction.

(vi) By (v), G has no perfect matching containing by. So we just need to consider perfect matchings in $G - by$. It is clear that a perfect matching of G contains exactly one of $\{aw, bx\}$ or $\{ax, bw\}$, and exactly one of $\{cy, dz\}$ or $\{cz, dy\}$. Thus, the number of perfect matchings $= 2 \times 2 = 4$. $\square$

Problem 8. *For $n \geq 3$, find the number of perfect matchings of the cycle C_n.*

Solution. A perfect matching of C_n has exactly $n/2$ edges. Thus, if n is odd, there are no perfect matchings.

Suppose n is even. Let C_n be $v_1 v_2 \cdots v_n v_1$ and let M be a perfect matching of C_n. If $v_1 v_2$ is in M, then $v_i v_{i+1}$, for all odd i, is in M. On the other hand, if $v_1 v_2$ is not in M, then $v_i v_{i+1}$, for all even i, is in M. Thus, there are exactly two perfect matchings for C_n when n is even. $\square$

Problem 9. *For $n \geq 1$, find the number of perfect matchings of the graph $K(n, n)$.*

Solution. Let X and Y be the two partite sets of $K(n, n)$ with $X = \{x_1, x_2, \cdots, x_n\}$. For x_1, to form an edge $x_1 u_1$ in a perfect matching, there are n choices for u_1, i.e., $u_1 \in Y$. Next, for x_2, to form the edge $x_2 u_2$, there are $n - 1$ choices for u_2, i.e., $u_2 \in Y \setminus \{u_1\}$. Continuing in this manner, we see that there are $n!$ perfect matchings for $K(n, n)$. $\square$

Problem 10. (+) *For $n \geq 2$, find the number of perfect matchings of the graph K_n.*

Solution. A perfect matching of K_n has exactly $\frac{n}{2}$ edges. Thus, if n is odd, there are no perfect matchings.

Suppose n is even. Choose a vertex u in $V(K_n)$. Then u has $n-1$ choices to choose a vertex, say v, to match with. Next, choose a vertex w in $V(K_n) \setminus \{u, v\}$. Then w has $n-3$ choices to choose a vertex to match with. Continuing this procedure, we see that the number of perfect matchings in K_n is

$$(n-1)(n-3)\cdots 3 \cdot 1.$$

$\square$

Problem 11. *Consider the following bipartite graph G with bipartition (X, Y):*

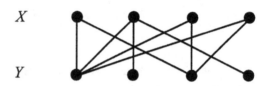

(i) *Does G contain a complete matching from X to Y?*

(ii) *Does G contain a perfect matching?*

(iii) *Find two matchings M and M' in G with $|M| = |M'| = 3$.*

(iv) *How many matchings M are there in G with $|M| = 3$?*

Solution. Let the vertices of X be, from left to right in the figure, x_1, x_2, x_3, x_4. Let the vertices of Y be, from left to right in the figure, y_1, y_2, y_3, y_4.

(i) No. Observe that y_2 and y_4 are adjacent only to x_2. Thus, there is no complete matching from Y to X. Since $|Y| = |X|$, there is also no complete matching from X to Y.

(ii) No. Since G does not contain a complete matching from X to Y, it does not contain a perfect matching.

(iii) $M = \{x_1y_1, x_2y_2, x_3y_3\}$; $M' = \{x_1y_1, x_2y_2, x_4y_3\}$.

(iv) Observe that in any matching M in G with $|M| = 3$, we must have either the edge x_2y_2 or the edge x_2y_4. When we remove the 3 vertices x_2, y_2 and y_4 from G, we are left with a complete bipartite graph $K(3,2)$. The number of matchings M' in $K(3,2)$ with $|M'| = 2$ is 3×2. Thus, there are $2 \times 3 \times 2 = 12$ such matchings. □

Problem 12. (+) *Consider the following 6×6 grid-board whose upper left and lower right corner squares are removed. You are given 17 dominoes, each covering exactly two adjacent squares (squares that have an edge in common) of the board. Can you use them to cover the 34 squares in the board?*

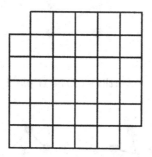

Solution. Colour the 34 squares black and white as shown below:

Clearly, each domino covers two squares with different colours. Form a bipartite graph G with bipartition (X, Y), where X is the set of black squares and Y is the set of white squares, and a vertex in X is adjacent to a vertex in Y if and only if their corresponding squares can be covered by a domino. The problem is equivalent to asking whether G contains a perfect matching. The answer is 'no' as $|X| = 18$ and $|Y| = 16$. □

Problem 13. (+) *Prove that the following 3-regular graph (triple flyswat) does not have a perfect matching, but does have a matching with seven edges.*

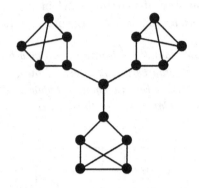

Solution. Let us label the vertices of the triple flyswat G as follows:

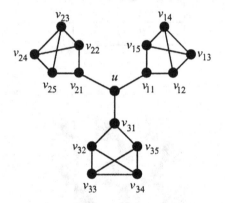

Suppose the graph G has a perfect matching M. We may assume that $uv_{11} \in M$. Now we observe that the 5 vertices in $W = \{v_{21}, v_{22}, v_{23}, v_{24}, v_{25}\}$ can only be incident to edges of M which have both end-vertices in W. This is however impossible as there is an odd number, 5, of vertices in W. Thus, G does not have a perfect matching.

A matching with seven edges is

$$\{uv_{11}, v_{12}v_{13}, v_{14}v_{15}, v_{22}v_{23}, v_{24}v_{25}, v_{32}v_{33}, v_{34}v_{35}\}.$$

$\square$

Problem 14. *Let T be a tree of order $n \geq 2$ and M, a maximal matching in T.*

(i) *What is the largest possible value for $|M|$? Construct one such T which contains one such M having its $|M|$ attaining this largest value.*

(ii) *What is the least possible value for $|M|$? Construct one such T which contains one such M having its $|M|$ attaining this least value.*

Solution. (i) Clearly, $|M| \leq \lfloor \frac{n}{2} \rfloor$.

Consider the path $P_n : v_1 v_1 v_2 \cdots v_n$. Then
$$M' = \left\{ v_i v_{i+1} \,\middle|\, i = 1, 3, 5, \cdots, 2 \left\lfloor \frac{n}{2} \right\rfloor - 1 \right\}$$
is a matching with $|M'| = \lfloor \frac{n}{2} \rfloor$.

Thus, the largest possible value for $|M|$ is $\lfloor \frac{n}{2} \rfloor$.

(ii) Clearly, $|M| \geq 1$.

Consider S_n, the star of order n, which is also a tree. (Recall that a star of order n has one vertex of degree $n - 1$, and $n - 1$ end-vertices.) Clearly, any matching in S_n contains exactly one edge.

Thus, the least possible value for $|M|$ is 1. $\qquad\square$

Problem 15. ($*$) *The twenty members of a local tennis club have scheduled exactly 14 two-person games among themselves, with each member playing in at least one game. Prove that within this schedule there must be a set of 6 games with twelve distinct players. (USAMO, 1989)*

Solution. Represent the situation as a graph G with 20 vertices representing the 20 players and two vertices are adjacent if and only if a game is scheduled between the two corresponding players. Thus, there are 14 edges and the degree of each vertex is at least 1.

Let M be a matching in G such that $|M|$ is the largest among all the matchings in G. Let U be the set of vertices in G which are not incident with any edge in M. Then $|U| = 20 - 2|M|$, as shown below.

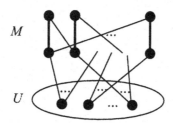

Note that

(1) each vertex in U is incident with an edge in G and

(2) no two vertices in U are joined by an edge in G (why?).

Thus, there are at least $|M| + |U|$ edges in G, and we have:

$$14 = e(G) \geq |M| + |U| = |M| + 20 - 2|M|,$$

which implies that $|M| \geq 6$, as desired.

Thus within this schedule there must be a set of 6 games with twelve distinct players.   □

Problem 16. (∗) *Generalize the result in the preceding problem by replacing 'twenty members' by 'n members', and '14 games' by 'm games'.*

Solution. The n members of a local tennis club have scheduled exactly m two-person games among themselves, with each member playing in at least one game. We claim that within this schedule there must be a set of $n - m$ games with $2(n - m)$ distinct players.

Represent the situation as a graph G with n vertices representing the n players and two vertices are adjacent if and only if a game is scheduled between the two corresponding players. Thus, there are m edges and the degree of each vertex is at least 1.

Let M be a matching in G such that $|M|$ is the largest among all the matchings in G. Let U be the set of vertices in G which are not incident with any edge in M. Then $|U| = n - 2|M|$. Note that

(1) each vertex in U is incident with an edge in G and

(2) no two vertices in U are joined by an edge in G.

Thus, there are at least $|M| + |U|$ edges in G, and we have:

$$m = e(G) \geq |M| + |U| = |M| + n - 2|M|,$$

which implies that $|M| \geq n - m$, as desired.

Thus within this schedule there must be a set of $n - m$ games with $2(n - m)$ distinct players.   □

Exercise 5.3

Problem 1. *Let G be the bipartite graph you constructed in Problem 1, Exercise 5.2. Let $S = \{A_1, A_3, A_5\}$. Find $N(S)$. Is $|N(S)| < |S|$? What conclusion can you draw from Theorem 5.1?*

Solution. We find that

$$N(S) = \{J_2, J_3\}.$$

So $|N(S)| = 2 < 3 = |S|$. By Theorem 5.1, the bipartite graph that models the situation in Problem 1 of Exercise 5.2 has no complete matchings. Thus it is impossible to assign each applicant a job for which he/she applies. $\square$

Problem 2. *Let G be the bipartite graph with bipartition (X, Y) as shown in Problem 7, Exercise 5.2.*

(i) *Complete the following table:*

S	$N(S)$
$\{a\}$	
$\{b\}$	
$\{c\}$	
$\{d\}$	
$\{a, b\}$	
$\{a, c\}$	
$\{a, d\}$	
$\{b, c\}$	
$\{b, d\}$	
$\{c, d\}$	
$\{a, b, c\}$	
$\{a, b, d\}$	
$\{a, c, d\}$	
$\{b, c, d\}$	
X	

(ii) *Is it true that $|S| \le |N(S)|$ for all $S \subseteq X$?*

(iii) *What conclusion can you draw from Corollary 5.2?*

Solution.   (i)

S	$N(S)$
$\{a\}$	$\{w, x\}$
$\{b\}$	$\{w, x, y\}$
$\{c\}$	$\{y, z\}$
$\{d\}$	$\{y, z\}$
$\{a, b\}$	$\{w, x, y\}$
$\{a, c\}$	$\{w, x, y, z\}$
$\{a, d\}$	$\{w, x, y, z\}$
$\{b, c\}$	$\{w, x, y, z\}$
$\{b, d\}$	$\{w, x, y, z\}$
$\{c, d\}$	$\{y, z\}$
$\{a, b, c\}$	$\{w, x, y, z\}$
$\{a, b, d\}$	$\{w, x, y, z\}$
$\{a, c, d\}$	$\{w, x, y, z\}$
$\{b, c, d\}$	$\{w, x, y, z\}$
X	$\{w, x, y, z\}$

(ii) It is true.

(iii) Since $|S| \leq |N(S)|$ for every subset S of X and $|X| = |Y|$, by Corollary 5.2, the bipartite graph shown in Problem 7, Exercise 5.2, has a perfect matching.   □

Problem 3. *Let G be the bipartite graph with bipartition (X, Y) as shown in Problem 11, Exercise 5.2.*

(i) *Find a subset S of X such that $|S| > |N(S)|$.*

(ii) *What conclusion can you draw from Theorem 5.1?*

Solution.   (i) We first label the vertices in the bipartite graph in Problem 11, Exercise 5.2 as follows:

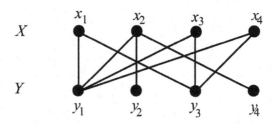

Let $S = \{x_1, x_3, x_4\}$. Note that $N(S) = \{y_1, y_3\}$ and $|S| > |N(S)|$.

(ii) By Theorem 5.1, the bipartite graph in Problem 11, Exercise 5.2 has no complete matchings from X to Y. ☐

Problem 4. *Consider the following bipartite graph G with bipartition* (X, Y):

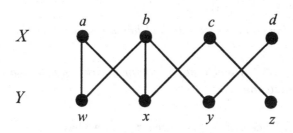

(i) *Let $S = \{a, b\}$. Find $N(S)$.*

(ii) *Let E_1 be the set of edges in G incident with some vertex in S. Find E_1.*

(iii) *Let E_2 be the set of edges in G incident with some vertex in $N(S)$. Find E_2.*

(iv) *Is $E_1 \subseteq E_2$?*

Solution. (i) $N(S) = \{w, x, y\}$.

(ii) $E_1 = \{aw, ax, bw, bx, by\}$.

(iii) $E_2 = \{aw, bw, ax, bx, cx, by, dy\}$.

(iv) Yes, it is true that $E_1 \subseteq E_2$. ☐

Problem 5. *Let G be a bipartite graph with bipartition (X, Y). For $S \subseteq X$, let E_1 be the set of edges in G incident with some vertex in S, and let E_2 be the set of edges in G incident with some vertex in $N(S)$. Is it true in general that $E_1 \subseteq E_2$? Why?*

Solution. It is true in general that $E_1 \subseteq E_2$.

Let $xy \in E_1$, where $x \in X$ and $y \in Y$. Then $x \in S$ by the definition of E_1. By the definition of $N(S)$, $y \in N(S)$. Thus $xy \in E_2$.

Hence $E_1 \subseteq E_2$. ☐

Problem 6. $(+)$ *Let G be a bipartite graph with bipartition (X, Y). Assume that there exists a positive integer k such that $d(y) \leq k \leq d(x)$ for each vertex y in Y and each vertex x in X.*

Let $S \subseteq X$, and denote by E_1 the set of edges in G incident with some vertex in S, and by E_2 the set of edges in G incident with some vertex in $N(S)$ (see the preceding problem).

(i) *Show that $k|S| \leq |E_1| \leq |E_2| \leq k|N(S)|$.*

(ii) *Deduce from Theorem 5.1 that G has a complete matching from X to Y.*

(iii) *Deduce from (ii) that every k-regular bipartite graph with $k \geq 1$ has a perfect matching.*

Solution. (i) Since $d(x) \geq k$ for each $x \in X$, we have

$$|E_1| = \sum_{x \in S} d(x) \geq \sum_{x \in S} k = k|S|.$$

Since $d(y) \leq k$ for each $y \in Y$, we have

$$|E_2| = \sum_{y \in N(S)} d(y) \leq \sum_{y \in N(S)} k = k|N(S)|.$$

By the result of Problem 5, $|E_1| \leq |E_2|$. Thus we have

$$k|S| \leq |E_1| \leq |E_2| \leq k|N(S)|.$$

(ii) By (i), we have $|S| \leq |N(S)|$ for every subset S of X. Thus, by Theorem 5.1, G has a complete matching from X to Y.

(iii) Let G be a k-regular bipartite graph with bipartition (X, Y), where $k \geq 1$. By the result of Problem 4 in Exercise 3.1, we have $|X| = |Y|$. Thus, by (ii), G has a perfect matching. $\qquad\square$

Problem 7. $(*)$ *Let G be a bipartite graph with bipartition (X, Y). Let $\Delta(G) = k \geq 1$ and $X^* = \{x \in X \mid d(x) = k\}$. Assume that X^* is not empty. Determine whether the following statement $(\#)$ is true and justify your answer.*

$(\#)$ G contains a matching M such that every vertex in X^* is incident with an edge in M.

Solution. The statement $(\#)$ is true.

Let H be the subgraph of G induced by $X^* \cup N(X^*)$. Then $(X^*, N(X^*))$ is a bipartition of H.

Note that $d_H(x) = k \geq d_H(y)$ for each $x \in X^*$ and each $y \in N(X^*)$. By the result of part (ii) of Problem 6, H contains a complete matching M from X^* to $N(X^*)$. It is clear that M is also a matching of G, and every vertex in X^* is incident with an edge in M. $\qquad\square$

Problem 8. (∗) *Let G be a bipartite graph with bipartition (X, Y). Prove that G contains a complete matching from X to Y if and only if*

$$|X \backslash N(T)| \leq |Y \backslash T|$$

for all $T \subseteq Y$.

Solution. ($\Rightarrow$) Assume that G contains a complete matching from X to Y. Then, by Theorem 5.1, $|S| \leq |N(S)|$ for each $S \subseteq X$.

Let $T \subseteq Y$ and $S = X \backslash N(T)$.

Let $y \in N(S)$. There exists $x \in S$ such that $xy \in E(G)$. If $y \in T$, then $x \in N(T)$, implying that $x \notin S$, a contradiction. Thus $y \notin T$. Hence $N(S) \subseteq Y \backslash T$.

Therefore

$$|X \backslash N(T)| = |S| \leq |N(S)| \leq |Y \backslash T|.$$

($\Leftarrow$) Assume that $|X \backslash N(T)| \leq |Y \backslash T|$ for all $T \subseteq Y$.

Let $S \subseteq X$ and $T = Y \backslash N(S)$. So $Y \backslash T = N(S)$. Then $N(T) \subseteq X \backslash S$, implying that $S \subseteq X \backslash N(T)$. By the assumption

$$|X \backslash N(T)| \leq |Y \backslash T|,$$

we have

$$|S| \leq |X \backslash N(T)| \leq |Y \backslash T| = |N(S)|.$$

By Theorem 5.1, G contains a complete matching from X to Y. $\qquad\square$

Problem 9. *Let G be a bipartite graph. Prove that G contains a perfect matching if and only if $|S| \leq |N(S)|$ for all $S \subseteq V(G)$.*

Solution. Let (X, Y) be a bipartition of G.

($\Rightarrow$) Since G contains a perfect matching, G contains a complete matching from X to Y. By Theorem 5.1, $|S| \leq |N(S)|$ for all $S \subseteq X$.

Similarly, G contains a complete matching from Y to X. By Theorem 5.1, $|S| \leq |N(S)|$ for all $S \subseteq Y$.

Let $S \subseteq V(G)$. So $S = S_1 \cup S_2$, where $S_1 \subseteq X$ and $S_2 \subseteq Y$. Then

$$|S| = |S_1| + |S_2| \leq |N(S_1)| + |N(S_2)| = |N(S)|.$$

($\Leftarrow$) Assume that $|S| \leq |N(S)|$ for all $S \subseteq V(G)$. Then $|S| \leq |N(S)|$ for all $S \subseteq X$ and $|S| \leq |N(S)|$ for all $S \subseteq Y$. By Theorem 5.1, G contains a complete matching from X to Y, and also a complete matching from Y to X. Thus $|X| = |Y|$, implying these complete matchings are perfect. $\square$

Problem 10. ($*$) *Let G be a connected bipartite graph with bipartition (X, Y), where $|X| \geq 2$ and $|Y| \geq 2$. Prove that the following statements are equivalent:*

(i) *Each edge of G is contained in a perfect matching of G.*

(ii) *$|X| = |Y|$ and $|S| < |N(S)|$ for all $S \subset X$ with $S \neq \emptyset$.*

(iii) *$G - \{x, y\}$ has a perfect matching for any $x \in X$ and $y \in Y$.*

Solution.   (i) $\Rightarrow$ (ii)

Since G contains perfect matchings, we have $|X| = |Y|$, and by Theorem 5.1, $|S| \leq |N(S)|$ for all $S \subset X$.

Suppose that there exists $S \subset X$ with $S \neq \emptyset$ such that $|S| = |N(S)|$.

Since G is connected, there exists an edge xy with $x \in X \backslash S$ and $y \in N(S)$.

By (i), xy is contained in a perfect matching M of G. But then the vertices in S are matched with the vertices in $N(S) \setminus \{y\}$ under M, which is impossible as $|S| > |N(S) \setminus \{y\}|$.

(ii) $\Rightarrow$ (iii)

Let $x \in X$ and $y \in Y$ and $H = G - \{x, y\}$. Note that $(X \backslash \{x\}, Y \backslash \{y\})$ is a bipartition of H.

Let S be any set with $S \neq \emptyset$ and $S \subseteq X \backslash \{x\}$. We have $|N_H(S)| \geq |N_G(S)| - 1$. As $S \subset X$, by (ii), we have $|N_G(S)| \geq |S| + 1$. Thus

$$|N_H(S)| \geq |N_G(S)| - 1 \geq |S| + 1 - 1 = |S|.$$

By Theorem 5.1, H has a complete matching from $X \backslash \{x\}$ to $Y \backslash \{y\}$. Since $|X| = |Y|$ by (ii), this complete matching is perfect.

(iii) $\Rightarrow$ (i)

Let xy be any edge in G, where $x \in X$ and $y \in Y$.

By (iii), $G - \{x, y\}$ has a perfect matching M. Clearly, $M \cup \{xy\}$ is a perfect matching of G. $\square$

Problem 11. (∗) *Let G be a bipartite graph with bipartition (X, Y) such that $|X| - |Y| = p \geq 1$. Form a larger bipartite graph G^* with bipartition $(X, Y \cup Y^*)$, where $|Y^*| = p$, such that*

(1) G^* contains G as an induced subgraph, and

(2) every vertex in Y^* is adjacent to every vertex in X.

Prove that G^* has a perfect matching if and only if G has a matching with $|Y|$ edges.

Solution. (⇒) Suppose that G^* has a perfect matching M. Let

$$M' = M \setminus \{e \in M \mid e \text{ is incident with a vertex } y \in Y^*\}.$$

Then M' is a matching of G with $|M'| = |M| - |Y^*| = |X| - p = |Y|$.

(⇐) Assume that G has a matching M' with $|Y|$ edges.

Let $X' = \{x \in X \mid x \text{ is not incident with any edge in } M'\}$. Then $|X'| = p$ and $[X' \cup Y^*] \cong K(p, p)$. Clearly, $[X' \cup Y^*]$ possesses a perfect matching, say M''. Then $M' \cup M''$ is a perfect matching of G^*. □

Problem 12. (∗) *Let G be a bipartite graph with bipartition (X, Y), and let k be an integer such that $1 \leq k \leq |X|$. Show that G contains a matching M with $|M| = k$ if and only if*

$$|S| \leq |N(S)| + |X| - k$$

for all $S \subseteq X$.

Solution. By Theorem 5.1, the result holds if $k = |X|$. In the following we assume that $1 \leq k < |X|$.

Construct the bipartite graph G^* with bipartition $(X, Y \cup Y^*)$ such that G is an induced subgraph of G^* and

(1) $|Y^*| = |X| - k$;

(2) every vertex in Y^* is adjacent to every vertex in X.

(⇒) Assume that G contains a matching M with $|M| = k$.

Let

$$X' = \{x \in X \mid x \text{ is not incident with any edge in } M\}.$$

Clearly, $|X'| = |X| - k = |Y^*|$, and by the definition of G^*, $[X' \cup Y^*]$ is a complete bipartite graph with bipartition (X', Y^*). Thus $[X' \cup Y^*]$ contains a perfect matching, say M'.

It follows that $M \cup M'$ is a complete matching of G^* from X to $Y \cup Y^*$. By Theorem 5.1, for all $S \subseteq X$,

$$|S| \leq |N_{G^*}(S)| = |N_G(S)| + |Y^*| = |N(S)| + |X| - k.$$

($\Leftarrow$) Assume that for all $S \subseteq X$,

$$|S| \leq |N(S)| + |X| - k.$$

As $N_{G^*}(S) = N(S) \cup Y^*$, we have

$$|N_{G^*}(S)| = |N(S)| + |Y^*| = |N(S)| + |X| - k.$$

Thus, for all $S \subseteq X$,

$$|S| \leq |N_{G^*}(S)|.$$

By Theorem 5.1, G^* has a complete matching M^* from X to $Y \cup Y^*$. Let

$$M = M^* \backslash \{e \in M^* \mid e \text{ is incident with a vertex in } Y^*\}.$$

Then M is a matching of G with

$$|M| = |M^*| - |Y^*| = |X| - (|X| - k) = k.$$

This completes the proof. $\square$

Problem 13. (*) *Let G be a bipartite graph with bipartition (X, Y). Assume that $d(x) \geq 6$ and $d(y) \leq 8$ for all $x \in X$ and $y \in Y$. Prove that G contains a matching M with $|M| \geq \frac{3}{4}|X|$.*

Solution. By the result of Problem 12, we need only to show that for all $S \subseteq X$,

$$|S| \leq |N(S)| + |X| - \left\lceil \frac{3}{4}|X| \right\rceil.$$

If $|S| \leq |N(S)| + |X| - \frac{3}{4}|X|$, then

$$|S| \leq \left\lfloor |N(S)| + |X| - \frac{3}{4}|X| \right\rfloor = |N(S)| + |X| - \left\lceil \frac{3}{4}|X| \right\rceil.$$

Note that $|N(S)| + |X| - \frac{3}{4}|X| = |N(S)| + \frac{1}{4}|X|$. So it suffices to show that

$$|S| \leq |N(S)| + \frac{1}{4}|X|$$

for all $S \subseteq X$.

Let S be any subset of X and $H = [S \cup N(S)]$. Then $(S, N(S))$ is a bipartition of H. Note that

$$e(H) = \sum_{x \in S} d_H(x) = \sum_{x \in S} d_G(x) \geq 6|S|$$

and

$$e(H) = \sum_{y \in N(S)} d_H(y) \leq \sum_{y \in N(S)} d_G(y) \leq 8|N(S)|.$$

Thus

$$6|S| \leq 8|N(S)|,$$

i.e., $\frac{3}{4}|S| \leq |N(S)|$. Now, we have

$$|S| = \frac{3}{4}|S| + \frac{1}{4}|S| \leq \frac{3}{4}|S| + \frac{1}{4}|X| \leq |N(S)| + \frac{1}{4}|X|.$$

By the result of Problem 12, G contains a matching M with $|M| \geq \frac{3}{4}|X|$.
$\square$

Problem 14. (∗) *Let G be a bipartite graph with bipartition (X, Y). For $S \subseteq X$, the* **deficiency** $\rho(S)$ *of S is defined as*

$$\rho(S) = |S| - |N(S)|.$$

Assume that $d(x) \geq 3$ and $d(y) \leq 4$ for all $x \in X$ and $y \in Y$. Show that

$$|X| \geq 4\rho(S),$$

for all $S \subseteq X$.

Solution. Let S be any subset of X and $H = [S \cup N(S)]$. Then $(S, N(S))$ is a bipartition of H. Note that

$$e(H) = \sum_{x \in S} d_H(x) = \sum_{x \in S} d_G(x) \geq 3|S|$$

and

$$e(H) = \sum_{y \in N(S)} d_H(y) \leq \sum_{y \in N(S)} d_G(y) \leq 4|N(S)|.$$

Thus

$$4|N(S)| \geq 3|S|,$$

i.e., $|N(S)| \geq \frac{3}{4}|S|$, which implies that

$$\rho(S) = |S| - |N(S)| \leq |S| - \frac{3}{4}|S| = \frac{1}{4}|S| \leq \frac{1}{4}|X|.$$

It follows that $|X| \geq 4\rho(S)$ for all $S \subseteq X$.

$\square$

Problem 15. (∗) *Let k and n be positive integers with k ≤ n. A k × n matrix with entries from* $\{1, 2, \cdots, n\}$ *is called a* **Latin rectangle** *if each 'i' in* $\{1, 2, \cdots, n\}$ *appears exactly once in each row and at most once in each column. A k × n Latin rectangle is called a* **Latin square** *of order n if k = n.*

$$\text{Consider the } 3 \times 5 \text{ Latin rectangle } L = \begin{pmatrix} 1\,2\,3\,4\,5 \\ 5\,1\,2\,3\,4 \\ 4\,5\,1\,2\,3 \end{pmatrix}.$$

Define a bipartite graph G with bipartition (X, Y) associated with L as follows:

(i) $X = \{1, 2, 3, 4, 5\}$,

(ii) $Y = \{C_1, C_2, C_3, C_4, C_5\}$, *where* C_i *is the ith column of L,*

(iii) *'i' in X is adjacent to 'C_j' in Y if and only if 'i' does not appear in* 'C_j'.

(a) *Draw the diagram of G.*

(b) *What is the degree of each vertex in X? Why?*

(c) *What is the degree of each vertex in Y? Why?*

(d) *Is G 2-regular?*

(e) *Does G contain a perfect matching? Why?*

(f) *Display a perfect matching in G if your answer to (e) is 'YES'.*

(g) *Use the perfect matching obtained in (f) to append a new row to L to form a 4 × 5 Latin rectangle L'.*

(h) *Expand L' to form a Latin square of order 5.*

Solution. (a)

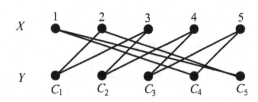

(b) $d(i) = 2$ for each $i \in X$, as every i does not appear in exactly 2 columns.

(c) $d(C_j) = 2$ for each $C_j \in Y$, as each column excludes exactly 2 members in X.

(d) Yes, G is 2-regular.

(e) Yes, G contains a perfect matching. By Corollary 5.3, every k-regular bipartite graph with $k \geq 1$ contains a perfect matching.

(f) The following is a perfect matching of G:

$$\{1C_5, 2C_1, 3C_2, 4C_3, 5C_4\}.$$

(g) From the above perfect matching, we get the following 4×5 Latin rectangle L':

$$L' = \begin{pmatrix} 1\,2\,3\,4\,5 \\ 5\,1\,2\,3\,4 \\ 4\,5\,1\,2\,3 \\ 2\,3\,4\,5\,1 \end{pmatrix}.$$

(h) We expand L' to form the following Latin square:

$$L'' = \begin{pmatrix} 1\,2\,3\,4\,5 \\ 5\,1\,2\,3\,4 \\ 4\,5\,1\,2\,3 \\ 2\,3\,4\,5\,1 \\ 3\,4\,5\,1\,2 \end{pmatrix}.$$

□

Problem 16. (∗) *Consider the* 2×6 *Latin rectangle* $L = \begin{pmatrix} 1\,2\,3\,4\,5\,6 \\ 3\,6\,4\,5\,2\,1 \end{pmatrix}$.

Define, likewise, the bipartite graph G *with bipartition* (X, Y) *associated with* L *as shown in the preceding problem by replacing* X *by* $\{1, 2, 3, 4, 5, 6\}$, *and* Y *by* $\{C_1, C_2, C_3, C_4, C_5, C_6\}$, *where* C_i *is the ith column of* L.

(a) Draw the diagram of G.

(b) What is the degree of each vertex in X? *Why?*

(c) What is the degree of each vertex in Y? *Why?*

(d) Is G *4-regular?*

(e) Does G *contain a perfect matching? Why?*

(f) Display a perfect matching in G *if your answer to (e) is 'YES'.*

(g) Use the perfect matching obtained in (f) to append a new row to L *to form a* 3×6 *Latin rectangle* L'.

Solution.   (a)

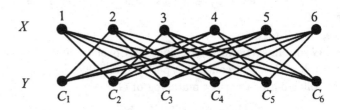

(b) $d(i) = 4$ for each $i \in X$, as every i does not appear in exactly 4 columns.

(c) $d(C_j) = 4$ for each $C_j \in Y$, as each column excludes exactly 4 numbers in X.

(d) Yes, G is 4-regular.

(e) Yes, G contains a perfect matching. Since G is 4-regular, by Corollary 5.3, it contains a perfect matching.

(f) The following is a perfect matching of G:

$$\{1C_4, 2C_6, 3C_2, 4C_5, 5C_3, 6C_1\}.$$

(g) From the above perfect matching, we get the following 3×6 Latin rectangle L':

$$L' = \begin{pmatrix} 1\ 2\ 3\ 4\ 5\ 6 \\ 3\ 6\ 4\ 5\ 2\ 1 \\ 6\ 3\ 5\ 1\ 4\ 2 \end{pmatrix}.$$

□

Exercise 5.4

Problem 1. *Find an SDR for each of the following families of sets.*

(i) $(\{1,2\}, \{2,5\}, \{4\}, \{3,5\})$;

(ii) $(\{1,3\}, \{1,3\}, \{2\}, \{1,4\}, \{4,5\})$;

(iii) $(\{5\}, \{1,6\}, \{2,3\}, \{1,4\}, \{3\}, \{1,4\})$.

Solution.
    (i) $(1,2,4,5)$.
    (ii) $(1,3,2,4,5)$.
    (iii) $(5,6,2,1,3,4)$.   □

Problem 2. *For each of the following families of sets, determine whether it has an SDR. Justify your answers.*

(i) $(\{1\}, \{2,3\}, \{1,2\}, \{1,3\}, \{1,4,5\})$;

(ii) $(\{1,2\}, \{2,3\}, \{4,5\}, \{4,5\})$;

(iii) $(\{1,2\}, \{2,3\}, \{3,4\}, \{4,5\}, \{5,1\})$.

Solution.   (i) Since $|\{1\} \cup \{2,3\} \cup \{1,2\} \cup \{1,3\}| = 3 < 4$, by Theorem 5.4, this family has no SDR.
    (ii) It has an SDR: $(1,2,4,5)$.
    (iii) It has an SDR: $(1,2,3,4,5)$.   □

Problem 3. *There are four clubs in a school with their committee members as shown below:*
    Club (A): $\{a,b\}$,
    Club (B): $\{a,c,e\}$,
    Club (C): $\{b,c\}$,
    Club (D): $\{b,d\}$.
    Let $X = \{A,B,C,D\}$ *and* $Y = \{a,b,c,d,e\}$. *Form a bipartite graph* G *with bipartition* (X,Y) *as follows: A vertex (club) in* X *is adjacent to a vertex (person) in* Y *if and only if that person is a committee member of the club.*

(i) *Draw the graph* G.

(ii) *Let* $S = \{A\}$. *Find* $N(S)$ *in* G.

(iii) *Let* $S = \{A,B\}$. *Find* $N(S)$ *in* G.

(iv) *Let $S = \{A, B, C\}$. Find $N(S)$ in G.*

(v) *Does there exist a subset S of X such that $|S| > |N(S)|$?*

(vi) *Does there exist a complete matching from X to Y?*

(vii) *Display a complete matching M from X to Y if your answer to (vi) is 'YES'.*

(viii) *Provide an SDR for the family (A, B, C, D) from M.*

Solution. (i) The graph G is shown below:

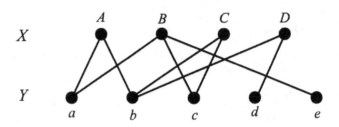

(ii) If $S = \{A\}$, then $N(S) = \{a, b\}$.
(iii) If $S = \{A, B\}$, then $N(S) = \{a, b, c, e\}$.
(iv) If $S = \{A, B, C\}$, then $N(S) = \{a, b, c, e\}$.
(v) There are no subsets S of X such that $|S| > |N(S)|$.
(vi) By Theorem 5.4, there exists a complete matching from X to Y.
(vii) The following is a complete matching M from X to Y:

$$M = \{Aa, Bc, Cb, Dd\}.$$

(viii) From M given in (vii), (A, B, C, D) has an SDR: (a, c, b, d). ☐

Problem 4. *Let $S_1 = \{b, c\}$, $S_2 = \{a\}$, $S_3 = \{a, b\}$ and $S_4 = \{c, d\}$. Verify that the family (S_1, S_2, S_3, S_4) satisfies the condition stated in Theorem 5.4, and thus conclude that the family has an SDR. Provide also one such SDR.*

Solution. Let $I \subseteq \{1, 2, 3, 4\}$. It is clear that if $|I| \leq 2$, then

$$\left| \bigcup_{i \in I} S_i \right| \geq |I|.$$

If $I = \{1, 2, 3\}$, then

$$\left| \bigcup_{i \in I} S_i \right| = |\{a, b, c\}| = 3 = |I|;$$

if $I = \{1, 2, 4\}$, then

$$\left| \bigcup_{i \in I} S_i \right| = |\{a, b, c, d\}| = 4 > |I|;$$

if $I = \{1, 3, 4\}$, then

$$\left| \bigcup_{i \in I} S_i \right| = |\{a, b, c, d\}| = 4 > |I|;$$

if $I = \{2, 3, 4\}$, then

$$\left| \bigcup_{i \in I} S_i \right| = |\{a, b, c, d\}| = 4 > |I|.$$

if $I = \{1, 2, 3, 4\}$, then

$$\left| \bigcup_{i \in I} S_i \right| = |\{a, b, c, d\}| = 4 = |I|.$$

So the family (S_1, S_2, S_3, S_4) satisfies the condition in Theorem 5.4. By Theorem 5.4, (S_1, S_2, S_3, S_4) contains an SDR.

Note that (c, a, b, d) is an SDR of (S_1, S_2, S_3, S_4). □

Problem 5. *Six teachers A, B, C, D, E and F are members of five committees. The memberships of the committees are*

$$\{A, B, C\}, \ \{D, E, F\}, \ \{A, D, E, F\}, \ \{A, C, E, F\} \ and \ \{A, B, F\}.$$

The activities of each committee are to be reviewed by a teacher who is not on the committee, and different committees are to be reviewed by different teachers. Can five distinct teachers be selected? If 'YES', show one such assignment.

Solution. Let S_i be the set of teachers who are not in committee i. Then we have

$$S_1 = \{D, E, F\},$$
$$S_2 = \{A, B, C\},$$
$$S_3 = \{B, C\},$$
$$S_4 = \{B, D\},$$
$$S_5 = \{C, D, E\}.$$

Observe that $(S_1, S_2, S_3, S_4, S_5)$ has an SDR: (E, A, B, D, C).

Hence five different teachers can be selected such that each teacher is not in the corresponding committee. □

Problem 6. *Show that each of the following families of sets has* **no** *SDR by Theorem 5.4.*

(i) $(\{1,2\},\{1\},\{3,4\},\{2\})$;

(ii) $(\{1\},\{2,3\},\{1,4,5\},\{1,2\},\{1,3\})$;

(iii) $(\{2,3\},\{2,3,4,5,6\},\{3,4\},\{4,5\},\{2,5\},\{2,4\})$.

Solution. (i) Since $|\{1,2\}\cup\{1\}\cup\{2\}| = 2 < 3$, by Theorem 5.4, this family has no SDR.

(ii) Since $|(\{1\}\cup\{2,3\}\cup\{1,2\}\cup\{1,3\}| = 3 < 4$, by Theorem 5.4, this family has no SDR.

(iii) Since $|(\{2,3\}\cup\{3,4\}\cup\{4,5\}\cup\{2,5\}\cup\{2,4\}| = 4 < 5$, by Theorem 5.4, this family has no SDR.   □

Problem 7. *For $n \geq 2$, let $S_1 = \{1\}$, $S_2 = \{1,2\}$ and for each $i = 3, \cdots, n$, let $S_i = \{1, 2, \cdots, i\}$.*

(i) *Show that the family $(S_1, S_2, \cdots, S_n)$ has an SDR.*

(ii) *How many different SDR's does $(S_1, S_2, \cdots, S_n)$ have?*

Solution. (i) Note that $(1, 2, 3, \cdots, n)$ is an SDR of $(S_1, S_2, \cdots, S_n)$.

(ii) We claim that $(S_1, S_2, \cdots, S_n)$ has only one SDR, i.e., $(1, 2, 3, \cdots, n)$. Suppose that $(S_1, S_2, \cdots, S_n)$ has an SDR $(x_1, x_2, \cdots, x_n)$. Then $x_1 \in S_1$ and so $x_1 = 1$; $x_2 \in S_2 \setminus \{1\}$ and so $x_2 = 2$. Likewise, we have $x_i = i$ for each $i = 3, \cdots, n$. Thus $(x_1, x_2, \cdots, x_n) = (1, 2, \cdots, n)$. □

Problem 8. *For $n \geq 2$ and for each $i = 1, 2, \cdots, n-1$, let $S_i = \{i, i+1\}$, and $S_n = \{n, 1\}$.*

(i) *Show that the family $(S_1, S_2, \cdots, S_n)$ has an SDR.*

(ii) *How many different SDR's does $(S_1, S_2, \cdots, S_n)$ have?*

Solution. (i) Since $i \in S_i$ for $i = 1, 2, \cdots n$, we have

$$\left| \bigcup_{i \in I} S_i \right| \geq |I|$$

for all $I \subseteq \{1, 2, \cdots, n\}$. By Theorem 5.4, $(S_1, S_2, \cdots, S_n)$ has an SDR.

(ii) We shall show that $(S_1, S_2, \cdots, S_n)$ has exactly two SDR's, namely,

$$(1, 2, \cdots, n) \quad \text{and} \quad (2, 3, \cdots, n, 1).$$

Let $(x_1, x_2, \cdots, x_n)$ be an SDR of $(S_1, S_2, \cdots, S_n)$.

Since $S_1 = \{1, 2\}$, we have either $x_1 = 1$ or $x_1 = 2$.

Case 1: $x_1 = 1$.

Then $x_n \neq 1$. As $S_n = \{n, 1\}$, we have $x_n = n$. Then $x_{n-1} \neq n$. As $S_{n-1} = \{n-1, n\}$, we have $x_{n-1} = n-1$. Continuing this argument, we have $x_i = i$ for each $i = n-2, n-3, \cdots, 2$. Thus $(x_1, x_2, \cdots, x_n) = (1, 2, \cdots, n)$.

Case 2: $x_1 = 2$.

Then $x_2 \neq 2$. As $S_2 = \{2, 3\}$, $x_2 = 3$. Then $x_3 \neq 3$. As $S_3 = \{3, 4\}$, $x_3 = 4$. Continuing this argument, we have $x_i = i + 1$ for each $i = 2, 3, \cdots, n-1$ and $x_n = 1$. Thus $(x_1, x_2, \cdots, x_n) = (2, 3, \cdots, n, 1)$. $\square$

Problem 9. (+) *Let* $S_1 = \{1, a\}$, $S_2 = \{1, 2a - 1\}$, $S_3 = \{2, 4 - a\}$ *and* $S_4 = \{2, a + 1\}$, *where* $a \in \{1, 2, 3, 4\}$. *Find all possible values of 'a' for which the family* (S_1, S_2, S_3, S_4) *has an SDR.*

Solution. If $a = 1$, then $S_1 = \{1\} = S_2$, and so (S_1, S_2, S_3, S_4) has no SDR.

If $a = 2$, then $S_1 = \{1, 2\}$, $S_2 = \{1, 3\}$, $S_3 = \{2\}$ and $S_4 = \{2, 3\}$. Since
$$|S_1 \cup S_2 \cup S_3 \cup S_4| = 3 < 4,$$
(S_1, S_2, S_3, S_4) has no SDR.

If $a = 3$, then $S_1 = \{1, 3\}$, $S_2 = \{1, 5\}$, $S_3 = \{2, 1\}$ and $S_4 = \{2, 4\}$. Observe that (S_1, S_2, S_3, S_4) has an SDR: $(1, 5, 2, 4)$.

If $a = 4$, then $S_1 = \{1, 4\}$, $S_2 = \{1, 7\}$, $S_3 = \{2, 0\}$ and $S_4 = \{2, 5\}$. Observe that (S_1, S_2, S_3, S_4) has an SDR: $(1, 7, 2, 5)$.

Hence (S_1, S_2, S_3, S_4) has an SDR if and only if $a \in \{3, 4\}$. $\square$

Problem 10. (+) *There are 12 clubs at a junior college. It is known that each club has at least 3 members and no student is a member of four or more clubs. Prove that this family of 12 clubs has an SDR.*

Solution. Let $X = \{S_1, S_2, \cdots, S_{12}\}$ be the set of 12 clubs and Y the set of students at the college. Let G be the bipartite graph with bipartition (X, Y) such that $S_i(\in X)$ is adjacent to $y(\in Y)$ if and only if $y \in S_i$.

By the given conditions, we have
$$d(y) \leq 3 \leq d(S_i)$$
for each $y \in Y$ and $i = 1, 2, \cdots, 12$.

Thus, by the result of Problem 6 (ii) in Exercise 5.3, G contains a complete matching M from X to Y. Let $M = \{S_1 y_1, S_2 y_2, \cdots, S_{12} y_{12}\}$. Then $(y_1, y_2, \cdots, y_{12})$ is an SDR of $(S_1, S_2, \cdots, S_{12})$. $\square$

Chapter 6

Eulerian Multigraphs and Hamiltonian Graphs

Theorem 6.1 *Let G be a connected multigraph. Then G is Eulerian if and only if every vertex in G is even.*

Theorem 6.2 *Let G be a connected multigraph. Then G is semi-Eulerian if and only if G has exactly two odd vertices. Moreover, if G is semi-Eulerian, then the two odd vertices in G are the initial and terminal vertices of any Euler trail in G.*

Theorem 6.3 *Let G be a graph. If G is Hamiltonian, then for any non-empty proper subset S of $V(G)$,*

$$c(G - S) \leq |S|.$$

Theorem 6.4 *Let G be a graph of order $n \geq 3$. If $d(v) \geq n/2$ for each vertex v in G, then G is Hamiltonian.*

Theorem 6.5 *Let G be a graph of order $n \geq 3$. If*

$$d(u) + d(v) \geq n$$

for every pair of non-adjacent vertices u and v in G, then G is Hamiltonian.

Exercise 6.1

Problem 1.   *Consider the following multigraph G of order 5.*

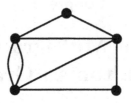

(i) *Find e(G).*

(ii) *Find in G a circuit with 2 edges; with 3 edges; with 4 edges.*

(iii) *Find in G a circuit with 5 edges that is not a cycle.*

(iv) *Find in G a circuit with 6 edges.*

(v) *If W is an Euler circuit in G, exactly how many edges are contained in W?*

(vi) *Does G contain an Euler circuit? Show one if there is.*

Solution.   (i) $e(G) = 8$.
   (ii) Name the vertices as shown below:

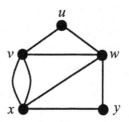

A circuit with 2 edges : xvx.
A circuit with 3 edges : $uvwu$.
A circuit with 4 edges : $xvwyx$.
(iii) The circuit $xvuwvx$ is an example.
(iv) The circuit $uvwxywu$ is an example.
(v) Eight ($= e(G)$) edges.
(vi) Yes, G contains an Euler circuit, e.g. $uvxvwxywu$.   □

Problem 2. *Five multigraphs are depicted below. Show that each of them is Eulerian by exhibiting an Euler circuit.*

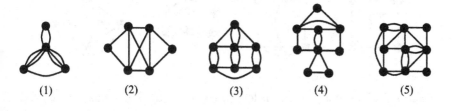

(1)   (2)   (3)   (4)   (5)

Solution. For each of the five Eulerian multigraphs, an Euler circuit is exhibited by labelling its edges as shown below.

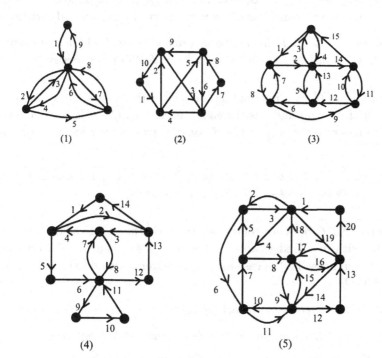

$\square$

Exercise 6.2

Problem 1.   *Determine whether the following multigraphs are Eulerian, semi-Eulerian or neither:*

Solution.   The first one (from left to right) is, by Theorems 6.1 and 6.2, neither Eulerian nor semi-Eulerian since it contains more than two odd vertices.

The second one is, by Theorem 6.2, semi-Eulerian as it contains exactly two odd vertices.

The third one is, by Theorem 6.1, Eulerian as it contains no odd vertices.

The last one is semi-Eulerian as it contains exactly two odd vertices. □

Problem 2.   *Let G be the multigraph considered in Problem 1 of Exercise 6.1. Does G contain a circuit with 7 edges? Justify your answer.*

Solution.   No, G does not contain a circuit with 7 edges. If G contains a circuit with 7 edges and e is the eighth edge, then $G - e$ is Eulerian. But this can never be the case as $G - f$ contains two odd vertices for any edge f in G. □

Problem 3.   *Let G be an Eulerian multigraph of size m. Can G contain a circuit with $m - 1$ edges? Justify your answer.*

Solution.   No, G does not contain a circuit with $m - 1$ edges. If G contains a circuit with $m - 1$ edges and e is the m-th edge, then $G - e$ is Eulerian. But this can never be the case as $G - f$ contains two odd vertices for any edge f in G. □

Problem 4.   *Determine if each of the following statements is true:*

(i) *If G is an Eulerian graph, then G contains no cut-vertices.*

(ii) *If G is an Eulerian graph, then G contains no bridges.*

(iii) *If G is an Eulerian graph of odd order and $\overline{G}$ is connected, then $\overline{G}$ is also an Eulerian graph.*

Solution. (i) Not true. For instance, the following Eulerian graph contains a cut-vertex.

(ii) True. If G contains a bridge, say uv, then $G-uv$ contains exactly two components. Let H be the component containing u. Since G is Eulerian, by Theorem 6.1, u is the only odd vertex in the graph H, which however contradicts Corollary 1.2.

(iii) True. Suppose that G is of order $2r+1$, for some positive integer r. As G is Eulerian, by Theorem 6.1, every vertex in G is even. Note that if $d(x) = 2k$ for a vertex x in G, then $d(x) = 2r - 2k$ in $\overline{G}$. Thus, every vertex is also even in $\overline{G}$. As $\overline{G}$ is assumed to be connected, $\overline{G}$ is Eulerian by Theorem 6.1. □

Problem 5. *Which $K(p,q)$, $p \geq q \geq 1$, are semi-Eulerian?*

Solution. By Theorem 6.2, $K(p,q)$ is semi-Eulerian when and only when $q = 2$ and p is odd.

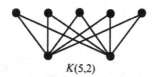

$K(5,2)$ □

Problem 6. *The following multigraph is semi-Eulerian as it contains exactly two odd vertices, namely, x and z.*

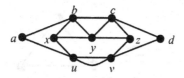

(i) *Form a new multigraph G^* by adding to G a new edge joining x and z. Is G^* Eulerian?*

(ii) *Find an Euler circuit W in G^*.*

(iii) *Delete xz (or zx) from W. Can you find an open Euler trail of G from the resulting sequence of edges?*

Solution.   (i) Yes, G^* is Eulerian since every vertex in G^* is even.

(ii) An Euler circuit W in G^* is given below:

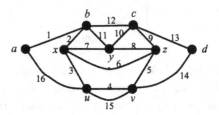

(iii) Yes, an open Euler $x-z$ trail of G can be obtained from the resulting sequence of edges after deleting 'zx' as shown below: $xyzcybcdvuabxuvz$ [namely, $(7)(8)(9)(10)(11)(12)(13)(14)(15)(16)(1)(2)(3)(4)(5)$].   □

Problem 7. (∗)   *Let G be a connected multigraph in which every vertex is even. Prove that G is Eulerian.*

Solution.   Let G be a connected multigraph in which every vertex is even. We shall prove that G is Eulerian by induction on $e(G)$ (≥ 2), the size of G.

When $e(G) = 2$, clearly,

$$G \;\cong\; \text{(graph of two vertices joined by two parallel edges)}$$

and when $e(G) = 3$,

$$G \;\cong\; \text{(triangle graph)}$$

and they are Eulerian.

Assume that the statement is true for all connected multigraphs G with $e(G) \leq k - 1$, where $k \geq 4$, in which every vertex is even.

Let G be a connected multigraph with $e(G) = k$ (≥ 4) in which every vertex is even. If G is 2-regular, then G is a cycle, and we are through.

Assume now that there is a vertex, say w, in G with $d(w) \geq 4$. Let e and f be two edges incident with w in G and let $H = G - \{e, f\}$.
Case 1. The edges e and f are parallel (see the diagram below).

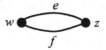

Clearly, every vertex in H is even (but H may not be connected).

(i) H is connected. As $e(H) < k$, by the induction hypothesis, H is Eulerian, and hence H possesses an Euler circuit W. Clearly, W can be extended to an Euler circuit W' of G by inserting e and f successively into W the moment the vertex w is visited when W is traversed.

(ii) H is disconnected. Then H has two components, say, H^* and H' (see the diagram below). By the induction hypothesis, both H^* and H' are Eulerian, and hence they possess Euler circuits W^* and W' respectively. Combine W^*, W' and $\{e, f\}$ as follows: starting at w and following W^* to return to w; traversing e to reach z and following W' to return to z; and finally, traversing f to terminate at w. This is clearly an Euler circuit of G.

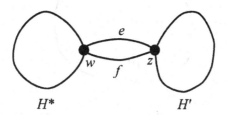

H^* $\qquad\qquad\qquad\qquad\qquad$ H'

Case 2. The edges e and f are non-parallel (see the diagram below, where $e = wx$ and $f = wy$).

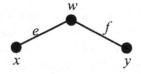

In this case, H has two odd vertices, namely, x and y. Let R be the multigraph obtained from H by adding a new edge joining x and y. Then every vertex in R is even (but R may not be connected).

(i) R is connected. As $e(R) < k$, by the induction hypothesis, R is Eulerian, and hence R possesses an Euler circuit W. Clearly, W can be converted to an Euler circuit of G if we replace the new edge joining x and y, say xy, in R by $e = xw$ and $f = wy$ successively.

(ii) R is disconnected. Then R has two components, say, R^* and R', where R^* contains w while R' contains the new edge xy (see the diagram below). By the induction hypothesis, both R^* and R' are Eulerian, and hence they possess Euler circuits W^* and W' respectively. We shall now combine $W^*, W' - xy$ and $\{e, f\}$ to produce an Euler circuit of G. We may assume that xy is the first edge in W'. Starting at x, we traverse $e(= xw)$ to reach w; following the whole W^* from w, we are back to w; traversing $f(= wy)$, we are now at y; finally, following the rest of $W' - xy$, we terminate at x.

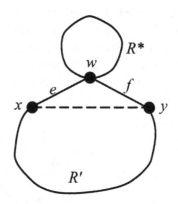

Clearly, the above closed walk is an Euler circuit of G. The proof is thus complete.   □

Problem 8. $(*)$ *Prove Theorem 6.2.*

Solution. $(\Rightarrow)$ Assume that G is semi-Eulerian. By definition, G possesses an open Euler trail W, say from x to y. Clearly, the trail W' obtained from W by adding yx at the end of W is an Euler circuit of $G + xy$ (the multigraph obtained from G by adding xy). Thus, $G + xy$ is Eulerian, and by Theorem 6.1, all vertices in $G + xy$ are even. It follows that x and y are the only two odd vertices in G.

$(\Leftarrow)$ Assume that G has exactly two odd vertices, say x and y. Then all vertices in $G + xy$ are even, and so $G + xy$ possesses an Euler circuit W. We may assume that xy is the first edge in traversing W. Then $W - xy$ forms an Euler $y - x$ trail of G (see the solution of Problem 6(iii) in this Exercise). This shows that G is semi-Eulerian. $\qquad \square$

Problem 9. *Two halls are partitioned into small rooms for an exhibition event in two different ways as shown in (a) and (b) below, where A is the entrance and B is the exit.*

(i) *Is it possible for a visitor to have a route which enters at A, passes through each door once and exactly once and exits at B in partition (a)?*

(ii) *Explain why such a route is not available in partition (b). Which door should be closed to ensure the existence of such a route?*

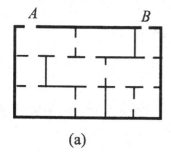

(a)

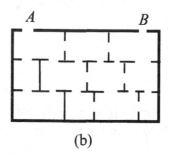

(b)

Solution. The situation can be modeled by a graph as follows: represent each of the rooms, including the outside area, by a vertex; and join two vertices by an edge if there is a door between the two corresponding rooms (or outside area). The resulting graphs of (a) and (b) are shown below:

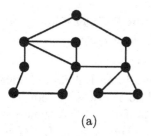

(a)

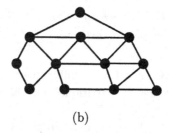

(b)

(i) The graph for (a) is Eulerian, and so the answer is YES.

(ii) The graph for (b) is non-Eulerian, and so such a route does not exist. If the door between the second and third rooms in row 2 is closed, then the degrees of the corresponding vertices reduce from 5 to 4, and the resulting graph becomes Eulerian.   □

Problem 10. (*)   *We have shown that the multigraph G of Figure 6.1 (a) (namely the multigraph below) is Eulerian.*

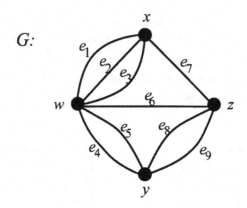

Look at its edge set $E(G)$ and observe that the edges in G can be partitioned into three edge-disjoint cycles as shown below:

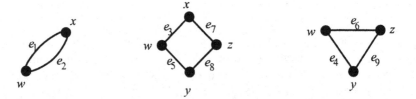

Show that, in general, a connected multigraph is Eulerian if and only if all its edges can be partitioned into some edge-disjoint cycles.

Solution. ($\Rightarrow$) We shall prove that the edges in an Eulerian multigraph can always be partitioned into some edge-disjoint cycles by induction on p, the number of cycles in G. It is obvious if $p = 1$. Assume that it holds for Eulerian multigraphs with $p < k$, where $k \geq 2$. Let G be an Eulerian multigraph with $p = k$. By Theorem 6.1, all vertices in G are even. Choose a cycle, say C, in G. Let H be the multigraph obtained from G by deleting the edges in C (note that H may be disconnected). Observe that each component of H has all its vertices even (and hence is Eulerian), and possesses less than k cycles. Thus, by the induction hypothesis, the edges in each component of H can be partitioned into edge-disjoint cycles. Accordingly, the edges in G can be partitioned into edge-disjoint cycles, in which C is a member.

($\Leftarrow$) Let G be a connected multigraph. Assume now that the edges in G can be partitioned into edge-disjoint cycles. We shall show that G is Eulerian. Let C be a cycle in this partition. If C includes all edges in G, then G is Eulerian. Otherwise, as G is connected, there is another cycle, say C', in this partition which has at least one vertex v in common with C. The (closed) walk P that starts at v and consists of the cycles C and C' in succession is a closed trail containing the edges of these two cycles. If P includes all edges of G, then G is Eulerian. Otherwise, applying a similar argument, P can be extended to a longer closed trail P' of G. Continuing this process, a closed trail containing all edges in G can eventually be obtained, which then shows that G is Eulerian. $\qquad\square$

Problem 11. *Let G_1 and G_2 be two semi-Eulerian multigraphs.*

(i) *Is it possible to form a semi-Eulerian multigraph by adding a new edge joining a vertex u in G_1 and a vertex v in G_2 as shown below? If the answer is 'yes', how can this be done?*

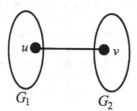

(ii) *Is it possible to form an Eulerian multigraph by adding two new edges, each of which joining a vertex in G_1 and a vertex in G_2? If the answer is 'yes', how can this be done?*

Solution. (i) Yes, join u and v, where u is an odd vertex in G_1 and v is an odd vertex in G_2. The resulting multigraph is now connected and has exactly two odd vertices. It is semi-Eulerian by Theorem 6.2.

(ii) Yes, let a and b (respectively, x and y) be the two odd vertices in G_1 (respectively, in G_2). Join a to x and b to y. The resulting multigraph is now connected and has no odd vertices. It is Eulerian by Theorem 6.1. ☐

Problem 12. (+)   *Let G_1 and G_2 be two connected multigraphs having $2p$ and $2q$ odd vertices respectively, where $1 \leq p \leq q$. We wish to form an Eulerian multigraph from G_1 and G_2 by adding new edges, each of which joining a vertex in G_1 and a vertex in G_2. What is the least number of edges that should be added? How can this be done?*

Solution. There are $2q$ odd vertices in G_2. To turn them into even in the resulting multigraph, it requires at least $2q$ new edges. The following construction shows that $2q$ new edges suffice. ☐

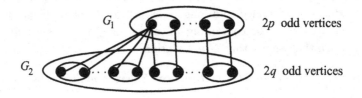

Problem 13. (+)   *The following graph H is not Eulerian. What is the least number of new edges that should be added to H so that the resulting multigraph becomes Eulerian? In how many ways can this be done?*

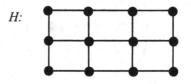

Solution. There are 6 odd vertices as shown in H.

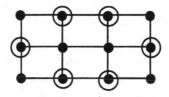

We pair them up into 3 pairs and join each pair by a new edge. Thus the least number of new edges is '3'. This can be done in $5 \times 3 \times 1$ ($= 15$) ways. □

Problem 14. *Let G be a semi-Eulerian multigraph of order 8 and size 18, and with $\delta(G) = 3$ and $\Delta(G) = 6$. Assume that G contains exactly two vertices of degree 6. How many vertices of degree 3 does G have? Justify your answer and construct one such multigraph.*

Solution. For $i = 3, 4, 5, 6$, let n_i denote the number of vertices of degree i in G. Then, by assumption,

$$n_3 + n_4 + n_5 + n_6 = 8, \ n_6 = 2 \text{ and } n_3 + n_5 = 2.$$

It follows that $n_4 = 4$. By Theorem 1.1,

$$3n_3 + 4 \times 4 + 5n_5 + 6 \times 2 = 2 \times 18,$$

that is,

$$3n_3 + 5n_5 = 8.$$

Solving $n_3 + n_5 = 2$ and $3n_3 + 5n_5 = 8$ yields $n_3 = 1$.

An example of G is shown below:

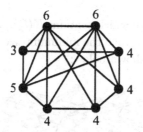

□

Problem 15. (+)   *Let G be a non-trivial connected multigraph. For $A \subset V(G)$, let $e(A, V(G)\backslash A)$ denote the number of edges in G having an end in A and the other in $V(G)\backslash A$ (see Problem 29 in Exercise 2.3). Show that G is Eulerian if and only if $e(A, V(G)\backslash A)$ is even for every proper subset A of $V(G)$.*

Solution. ($\Leftarrow$) By Theorem 6.1, we show that each vertex in G is even. Thus, let v be in $V(G)$ and $A = \{v\}$. Observe that $d(v) = e(A, V(G)\backslash A)$, which is even by assumption.

($\Rightarrow$) [First proof]

Let A be a proper subset of $V(G)$ and W an Euler circuit of G with the starting vertex v in A. Clearly, for each edge e in W from A to $V(G)\backslash A$, there is an edge e' in W from $V(G)\backslash A$ to A (as v is also the terminal vertex of W). It follows that $e(A, V(G)\backslash A)$ is even.

($\Rightarrow$) [Second proof]

For $A \subseteq V(G)$, we have (see the solution of Problem 29 in Exercise 2.3):

$$e(A, V(G)\backslash A) = \sum_{x \in A} d(x) - \sum_{x \in A} d_{[A]}(x)$$

where $d_{[A]}(x)$ denotes the degree of x in $[A]$. By assumption and Theorem 6.1, $\sum_{x \in A} d(x)$ is even, and by Theorem 1.1, $\sum_{x \in A} d_{[A]}(x)$ is even. It follows from the above equality that $e(A, V(G)\backslash A)$ is even. $\qquad\square$

Problem 16. (+) *Let G be a graph which contains $K(5, 6)$ as a spanning subgraph.*

(i) *If G is semi-Eulerian, find the minimum size of G, and construct one such extremal semi-Eulerian graph G.*

(ii) *If G is Eulerian, find the minimum size of G, and construct one such extremal Eulerian graph G.*

Solution. Let G be a graph which contains $K(5, 6)$ as a spanning subgraph.

(i) If G is semi-Eulerian, then the minimum size of G is 32 ($= 5 \times 6 + 2$). An example of such G is shown below:

(ii) If G is Eulerian, then the minimum size of G is 33 ($= 5 \times 6 + 3$). An example of such G is shown below:

$\qquad\square$

Problem 17. (+)   *Let G be a multigraph which contains $K(5,7)$ as a spanning subgraph.*

(i) *Assume that G is semi-Eulerian. Can G be simple? If 'YES', find the minimum size of G, and construct one such extremal semi-Eulerian graph G.*

(ii) *Assume that G is Eulerian. Prove that G cannot be simple. Find the minimum size of G, and construct one such extremal Eulerian multigraph G.*

Solution.   Let G be a multigraph which contains $K(5,7)$ as a spanning subgraph.

(i) Assume that G is semi-Eulerian.

Yes, G can be simple, and the minimum size of G is 40 ($= 5 \times 7 + 5$). An example of such G is shown below:

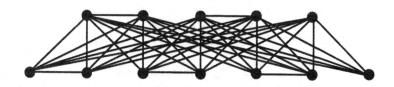

(ii) Assume that G is Eulerian. We first show that G cannot be simple. Suppose on the contrary that G is simple and let (X, Y) be the bipartition of $K(5,7)$ with $|X| = 5$. Since $d(x)$ is even in G, $d_{[X]}(x)$ ($= d(x) - 7$) is odd in $[X]$ for each x in X. Thus the graph $[X]$ consists of 5 odd vertices, which contradicts Corollary 1.2.

The minimum size of G is 41 ($= 5 \times 7 + 6$). An example of such G is shown below:

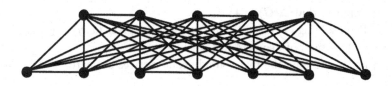

□

Problem 18. (+)   *Let G be an Eulerian graph of order 8 and size 10.*

(i) *Let k be the maximum possible value of $\Delta(G)$. Determine k and construct all such G with $\Delta(G) = k$.*

(ii) *Suppose that $\Delta(G) = 4$.*

   (a) Determine the number of vertices of degree 4 in G.

   (b) Assume further that no two vertices of degree 4 are adjacent. Construct all such G.

Solution.   Let G be an Eulerian graph of order 8 and size 10.

(i) Let k be the maximum possible value of $\Delta(G)$. As G is Eulerian, $\Delta(G)$ is even, and so $\Delta(G) \leq 6$. We claim that $k = 6$.

Let x, y and z be, respectively, the number of vertices of degree $2, 4$ and 6 in G. Then

$$x + y + z = 8 \tag{6.1}$$

and

$$2x + 4y + 6z = 20,$$

or

$$x + 2y + 3z = 10 \tag{6.2}$$

Solving the equations (6.1) and (6.2) yields one possible solution $(x, y, z) = (7, 0, 1)$. It follows that $k = 6$.

There is only one such G as shown below:

(ii) Suppose that $\Delta(G) = 4$.

(a) Then solving (1) and (2) in (a) with $z = 0$ gives $(x, y) = (6, 2)$. Thus there are 2 vertices of degree 4 in G.

(b) Let u and v be the two non-adjacent vertices of degree 4 in G. Note that

$$2 \leq |N(u) \cap N(v)| \leq 3.$$

Case 1. $|N(u) \cap N(v)| = 2$.

There are two such G's as shown below:

Case 2. $|N(u) \cap N(v)| = 3$.

There is only one such G as shown below:

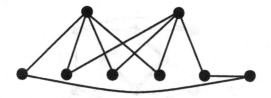

□

Exercise 6.4

Problem 1. *Determine whether the following graphs are Hamiltonian. Justify your answers.*

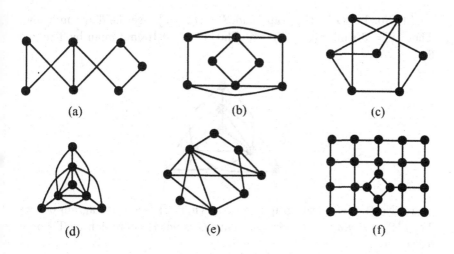

(a) (b) (c)

(d) (e) (f)

Solution.   (a) Yes. A spanning cycle is shown below:

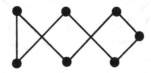

(b) No. Apply the method of 'degree two' (see rules for constructing Hamiltonian cycles) as shown below:

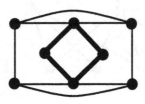

(c) Yes. A spanning cycle is shown below:

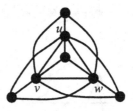

(d) No. Let G be the graph and $S = \{u, v, w\}$ (see the diagram below). Then $|S| = 3$ while $c(G - S) = 4$. Thus G is not Hamiltonian by Theorem 6.3.

(e) No. Let G be the graph and $S = \{u, v, w\}$ (see the diagram below). Then $|S| = 3$ while $c(G - S) = 4$. Thus G is not Hamiltonian by Theorem 6.3.

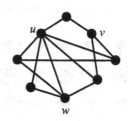

Note. You may also apply the method of 'degree two'.

    (f) Yes. A spanning cycle is shown below:

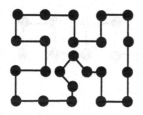

□

Problem 2. *Determine whether the following m × n rectangular grids are Hamiltonian.*

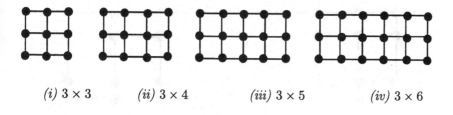

   (i) 3 × 3   *(ii)* 3 × 4   *(iii)* 3 × 5   *(iv)* 3 × 6

Solution.

    (i) No. Apply the method of 'degree two' or Theorem 6.3.

    (ii) Yes. A spanning cycle is shown below:

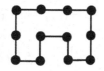

(iii) No. Apply Theorem 6.3 (see the diagram below).

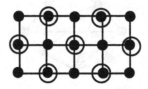

(iv) Yes. A spanning cycle is shown below:

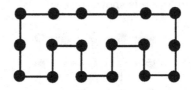

□

Problem 3. (+) *Show that an $m \times n$ rectangular grid is Hamiltonian if and only if either m or n is even.*

Solution. ($\Leftarrow$) We may assume that n is even. In this case, a spanning cycle is shown below:

($\Rightarrow$) Suppose that both m and n are odd. Consider G as the graph with

$$V(G) = \{(i,j) \mid i, j \text{ are integers}, 1 \leq i \leq n, 1 \leq j \leq m\}$$

and

$$E(G) = \{uv \mid u = (i,j) \text{ and } v = (i', j') \text{ in } V(G), |i - i'| + |j - j'| = 1\}.$$

Let

$$S = \{(i,j) \in V(G) \mid i + j \text{ is odd}\}$$

as shown below:

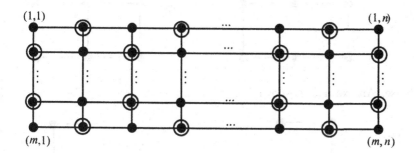

Observe that $|S| = \frac{mn-1}{2}$ and $c(G - S) = \frac{mn+1}{2} = |S| + 1$. Thus, G is not Hamiltonian by Theorem 6.3. □

Note: *Let G be a bipartite graph with bipartition (X, Y). If G is Hamiltonian, then $|X| = |Y|$ and hence the order of G is even.*

A proof of this fact can be found in the solution of Problem 6 of Exercise 3.1. Another proof by Theorem 6.3 is given below.

Suppose on the contrary that $|X| \neq |Y|$, say $|X| < |Y|$. We then observe that $c(G - X) = |Y| > |X|$. By Theorem 6.3, G is not Hamiltonian, a contradiction.

By the result of this note, we have a simple proof for the necessity of Problem 3.

Let G be the graph for the $m \times n$ rectangular grid. Notice that G is a bipartite graph of order mn. If G is Hamiltonian, then the order of G is even by the above note. Thus mn is even, implying that either m or n is even. □

Problem 4. *Let H be a spanning subgraph of a graph G. Which of the following statements is/are true?*

(i) *If H is Eulerian, then G is Eulerian.*

(ii) *If H is semi-Eulerian, then G is semi-Eulerian.*

(iii) *If H is Hamiltonian, then G is Hamiltonian.*

Solution. Let H be a spanning subgraph of G.

(i) False. An example is shown below:

H: 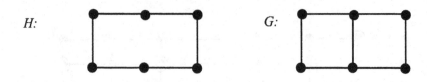 *G:*

(ii) False. An example is shown below:

H:  *G:*

(iii) True. Suppose H is Hamiltonian. Then H contains a spanning cycle C. Since $V(G) = V(H)$ and every edge in H is also in G, the cycle C is also a spanning cycle of G. Thus, G is Hamiltonian. □

Problem 5. *Prove that if G is a Hamiltonian graph, then $d(v) \geq 2$ for each vertex v in G.*

Solution. Suppose G is Hamiltonian. Then G contains a spanning cycle C. Thus, for each vertex v in G, $d_G(v) \geq d_C(v) = 2$. □

Problem 6.

(i) *Let H be a graph such that $d(v) = 2$ for each vertex v in H. Is H Hamiltonian?*

(ii) *Let H be a connected graph such that $d(v) = 2$ for each vertex v in H. Is H Hamiltonian?*

Solution. (i) No, as H may be disconnected. An example is shown below:

(ii) Yes. In this case, H is itself a cycle, and hence Hamiltonian. □

Problem 7. (+) *Let G be a Hamiltonian graph with a Hamiltonian cycle
C. For any non-empty proper subset A of $V(G)$, let $e_C(A, V(G)\backslash A)$ denote
the number of edges in C having an end in A and the other in $V(G)\backslash A$.
Show that $e_C(A, V(G)\backslash A)$ is always even.*

Solution. Let w be the first vertex in traversing the Hamiltonian cycle C
of G. We may assume that w is in A. Note that each time we leave A for
$V(G)\backslash A$ via an edge in C, as eventually we have to go back to w, there
must be another edge in C for us to leave $V(G)\backslash A$ for A. Thus all edges in
C linking A and $V(G)\backslash A$ are paired up, and the result follows. □

Problem 8.   *Consider the following graph G and let $S = \{x, y, z\}$.*

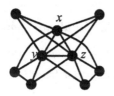

(i) *Draw the graph $G - S$.*

(ii) *Find $|S|$ and $c(G - S)$.*

(iii) *Is $|S| < c(G - S)$?*

(iv) *Is G Hamiltonian?*

Solution.   (i) The graph $G - S$ is shown below:

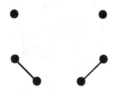

(ii) $|S| = 3$ while $c(G - S) = 4$.

(iii) Yes, $|S| < c(G - S)$.

(iv) By Theorem 6.3, G is not Hamiltonian. □

Problem 9.   *Let H be the graph depicted below:*

(i) *Verify that $c(H - S) \leq |S|$ for all non-empty proper subsets S of $V(H)$.*

(ii) *Is H Hamiltonian?*

(iii) *Is the converse of Theorem 6.3 true?*

Solution.   For convenience, name the vertices as shown below:

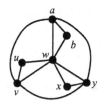

(i) As $v(H) = 7$, to verify that $c(H - S) \leq |S|$, it suffices to consider those subsets S of $V(H)$ with $1 \leq |S| \leq 3$. As H contains no cut-vertices, it is clear that $c(H - S) = 1$ when $|S| = 1$. For the rest, by the symmetric structure of H, we need only to check the cases as shown in the following table:

S	$|S|$	$c(H-S)$
$\{a,b\}$	2	1
$\{a,u\}$	2	1
$\{a,v\}$	2	1
$\{a,w\}$	2	2
$\{b,u\}$	2	1
$\{b,w\}$	2	1
$\{a,b,w\}$	3	1
$\{a,b,u\}$	3	1
$\{a,b,v\}$	3	1
$\{a,v,y\}$	3	1
$\{a,u,x\}$	3	1
$\{a,u,y\}$	3	1
$\{a,w,u\}$	3	2
$\{a,w,v\}$	3	3
$\{b,u,w\}$	3	1
$\{b,u,x\}$	3	1

Thus, the inequality holds.

(ii) No, H is not Hamiltonian (the reader may apply the method of 'degree two').

(iii) No, the above facts show that the converse of Theorem 6.3 is, in general, not true. □

Problem 10. *Let G be a Hamiltonian graph of order n and size m such that $m = n + 2$ and $n \geq 5$.*

(i) *Prove that $2 \leq d(x) \leq 4$ for each vertex x in G.*

(ii) *If G is also semi-Eulerian, what can be said about the structure of G?*

Solution. Let C be a Hamiltonian cycle of G.

(i) Let v be in $V(G)$. Clearly, $d(v) = d_G(v) \geq d_C(v) = 2$. On the other hand, as $m = n + 2$, v is incident with at most two edges not in C. Thus $d(v) \leq d_C(v) + 2 = 4$.

(ii) If G is semi-Eulerian, then G has exactly two odd vertices, and they must be of degree 3 by (i). The structure of G is thus of the following form:

☐

Problem 11. *Let G be a semi-Eulerian and Hamiltonian graph with $v(G) = 12$, $e(G) = 17$, $\delta(G) = 2$ and $\Delta(G) = 4$.*

(i) *How many vertices of degree 3 can G have?*

(ii) *How many vertices of degree 2 does G have?*

(iii) *Construct one such graph G.*

(iv) *Assume that the odd vertices are adjacent in G, but no two vertices of degree 2 are adjacent in G. What can be said about the structure of the subgraph induced by the set of vertices of degree 4 in G?*

Solution. (i) As G is semi-Eulerian, and $\delta(G) = 2$ and $\Delta(G) = 4$, G has two vertices of degree 3.

(ii) Let x be the number of vertices of degree 2 in G. Then

$$2x + 3 \times 2 + 4(10 - x) = 2 \times 17,$$

and so $x = 6$.

(iii) One such G is shown below:

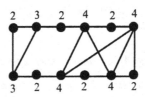

(iv) Let C be a Hamiltonian cycle of G as shown below:

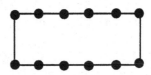

By assumption, we may assume that u, v, w, x, y and z are the six vertices of degree 2. Two of the remaining vertices must be adjacent to become vertices of degree 3. Thus the subgraph induced by the four vertices of degree 4 in G is a C_4 (see below).

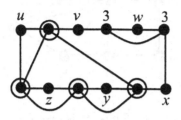

Problem 12. *Let H be a semi-Eulerian and Hamiltonian graph with a Hamiltonian cycle C. Assume that $v(H) = 7$, $e(H) = 12$, $\delta(H) = 2$ and $\Delta(H) = 5$, and that H has exactly 2 vertices of degree 2.*

(i) *Find the number of vertices of degree 4 and the number of vertices of degree 5 in H.*

(ii) *Assume that the 2 vertices of degree 2 are adjacent in C. Construct all such graphs H.*

Solution. (i) Let n_i denote the number of vertices of degree i in H. Then, by assumption,

$$n_3 + n_5 = 2, \quad n_2 = 2, \quad n_2 + n_3 + n_4 + n_5 = 7,$$

and

$$2n_2 + 3n_3 + 4n_4 + 5n_5 = 2 \times 12 = 24.$$

Solving these equations yields $(n_2, n_3, n_4, n_5) = (2, 1, 3, 1)$. Thus $n_4 = 3$ and $n_5 = 1$.

(ii) By assumption, H contains the following Hamiltonian cycle C in which the two adjacent vertices of degree 2 are indicated:

C:

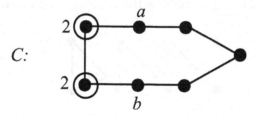

Clearly, either a or b (see the above diagram), say the former, is of degree 5, and we have:

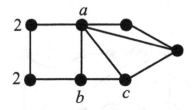

where $d(b) \neq 3$ and $d(c) \neq 3$. Thus, H is one of the following two (which are actually isomorphic):

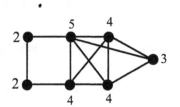   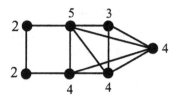

□

Problem 13.   (+) *Let G be a Hamiltonian bipartite graph of order 8.*

(i) *Explain why $\delta(G) \geq 2$ and $\Delta(G) \leq 4$.*

(ii) *Assume further that G is Eulerian and $\Delta(G) = 4$. What can be said about the structure of G?*

Solution.   (i) As G is Hamiltonian, $\delta(G) \geq 2$. Let (X, Y) be the bipartition of G. As G is Hamiltonian, $|X| = |Y| = 4$. Thus, $\Delta(G) \leq 4$.

(ii) Suppose that G contains a Hamiltonian cycle as shown below:

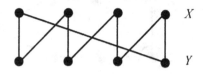

As $\Delta(G) = 4$, there is a vertex of degree 4 as shown below:

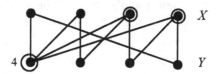

Since G is Eulerian, G contains no odd vertices. It follows that $G \cong K(4, 4)$, as shown in the following sequence of logical implications:

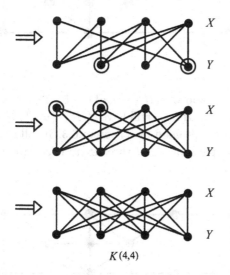

$K(4, 4)$

□

Problem 14. (+) *Let H be the graph given below:*

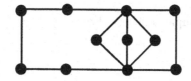

(i) *Is H Hamiltonian? Why?*

(ii) *Let $m(H)$ denote the minimum number of new edges that are needed to add to H so that the resulting graph H^* is Hamiltonian (note that $V(H^*) = V(H)$). Find the value of $m(H)$ and justify your answer.*

(iii) *Construct two such Hamiltonian graphs H^* obtained by adding $m(H)$ new edges to H.*

Solution.   (i) No, H is not Hamiltonian.   (The reader may apply the method of 'degree two' or Theorem 6.3 to justify it.)

(ii) $m(H) = 3$.

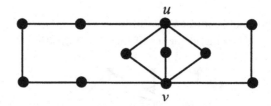

Note that, as shown above, $c(H - \{u, v\}) = 5$. Thus, by adding any two new edges to H to form H^*, we still have $c(H^* - \{u, v\}) \geq 3 > |\{u, v\}|$; that is, H^* is still non-Hamiltonian by Theorem 6.3. It follows that $m(H) \geq 3$. The following example shows that three extra edges are enough.

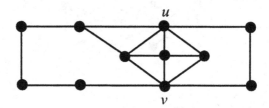

(iii) Another example is given below:

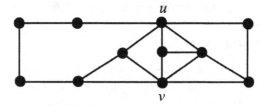

Problem 15. (+) *Let H be the graph given below:*

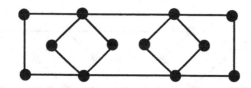

(i) *Is H Hamiltonian? Why?*

(ii) *Let m(H) denote the minimum number of new edges that are needed to add to H so that the resulting graph H* is Hamiltonian (note that V(H*) = V(H)). Find the value of m(H) and justify your answer.*

(iii) *Construct two such Hamiltonian graphs H* obtained by adding m(H) new edges to H.*

Solution. (i) No, H is not Hamiltonian. (The reader may apply the method of 'degree two' or Theorem 6.3 to justify it.)

(ii) $m(H) = 2$.

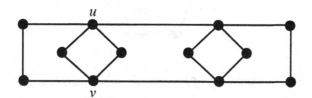

Note that, as shown above, $c(H - \{u, v\}) = 4$. Thus, by adding any one new edge to H to form H^*, we still have $c(H^* - \{u, v\}) \geq 3 > |\{u, v\}|$; that is, H^* is still non-Hamiltonian by Theorem 6.3. It follows that $m(H) \geq 2$. The following example shows that two extra edges are enough.

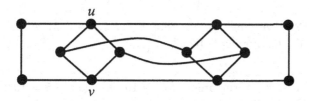

(iii) Another example is given below:

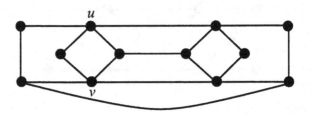

□

Exercise 6.5

Problem 1. *The following graph H is Hamiltonian.*

(i) *Does the Hamiltonicity of H follows from Theorem 6.4?*

(ii) *Does the Hamiltonicity of H follows from Theorem 6.5?*

Solution. Note that the order of H is 5.

(i) No, since $d(x) = 2 < 5/2$ for some vertex x in H.

(ii) Yes, because $d(x) + d(y) \geq 5$ for every two non-adjacent vertices x and y in H. □

Problem 2. *A graph G has $(8, 8, 8, 7, 7, 7, 6, 5, 5, 5)$ as its degree sequence. Is G Hamiltonian? Why?*

Solution. Yes. Since the order of G is 10 and $d(x) \geq 5 = 10/2$ for every vertex x in G, by Theorem 6.4, G is Hamiltonian. □

Problem 3. (+)   *Let G be a graph of order $n \geq 3$. The sufficient condition given by Dirac in Theorem 6.4 states that*

$$\textbf{(D)} \quad d(v) \geq n/2 \text{ for each } v \text{ in } V(G).$$

The sufficient condition given by Ore in Theorem 6.5 states that

(O) $d(u) + d(v) \geq n$ *for every pair of non-adjacent vertices u, v in $V(G)$.*

  (1) *Which of the following implications is true?*
  (i) **(D)** $\Rightarrow$ **(O)**;
  (ii) **(O)** $\Rightarrow$ **(D)**.
  (2) *Which of the following implications is true?*
  (i) *Theorem 6.4* $\Rightarrow$ *Theorem 6.5;*
  (ii) *Theorem 6.5* $\Rightarrow$ *Theorem 6.4.*

Solution.   (1) (i) is correct, since if (D) holds, then (O) holds.
  (2) (ii) is correct.   □

Problem 4. (+)   *Let G be a graph of order $n \geq 3$ and size m.*

(i) *Assume that there exist two non-adjacent vertices u and v in G such that $d(u) + d(v) \leq n - 1$. Show that $m \leq \binom{n-1}{2} + 1$.*

(ii) *Deduce that if $m \geq \binom{n-1}{2} + 2$, then G is Hamiltonian.*

Solution.   (i) Since u, v are not adjacent,

$$m = e(G - \{u, v\}) + d(u) + d(v) \leq \binom{n-2}{2} + n - 1 = \binom{n-1}{2} + 1.$$

  (ii) Since $m \geq \binom{n-1}{2} + 2$, by (i), we have $d(u) + d(v) \geq n$ for every two non-adjacent vertices u, v in G. Thus, by Theorem 6.5, G is Hamiltonian.
□

Problem 5.   *Construct a non-Hamiltonian graph of order $n \geq 3$ with size $\binom{n-1}{2} + 1$.*

Solution.   Let G be the following graph:

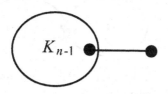

Clearly, $e(G) = \binom{n-1}{2} + 1$ and G is non-Hamiltonian.   □

Problem 6. (+)   *A path in a graph G is called a* **Hamiltonian path** *if it includes all the vertices in G.*

(i) *Is the following graph Hamiltonian? Does it contain a Hamiltonian path?*

(ii) *Prove that if G is a graph of order $n \geq 2$ such that $\delta(G) \geq (n-1)/2$, then G contains a Hamiltonian path.*

Solution.   (i) This graph is not Hamiltonian, but it contains a Hamiltonian path.

(ii) Let H be the graph obtained from G by adding one new vertex w and adding n edges joining w to every vertex in G.

Note that the order of H is $n+1$. For every vertex $x \in V(H)$, if $x = w$, then $d_H(x) = n$; otherwise,

$$d_H(x) = d_G(x) + 1 \geq \frac{n-1}{2} + 1 = \frac{n+1}{2}.$$

By Theorem 6.4, H contains a Hamiltonian cycle C. Observe that $C - w$ is a Hamiltonian path in G.   □

Problem 7. (+)   *For each odd integer $n \geq 3$, construct a non-Hamiltonian graph G of order n such that $\delta(G) = (n-1)/2$.*

Solution.   Let $n = 2k + 1$. Let G be the graph obtained from two K_{k+1}'s by gluing them at one vertex, denoted by w, as shown below:

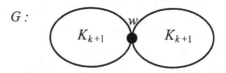

Note that the order of G is $2(k+1)-1 = 2k+1 = n$, and $d(x) = k = (n-1)/2$ for every vertex x in G, except w.

Since $G - w$ is disconnected, G is not Hamiltonian by Theorem 6.3. $\square$
Note: $K(k, k + 1)$ is another non-Hamiltonian graph G of order n such that $\delta(G) = (n - 1)/2$, where $n = 2k + 1$.

Problem 8. (+) *Let $G + H$ denote the **join** of two graphs G and H (see Problem 27 in Exercise 4.3). For a positive integer r, denote by rK_2 the union of r independent edges as shown below:*

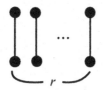

(i) *Determine whether the join $(3K_2) + N_7$ is Hamiltonian. Justify your answer.*

(ii) *Determine whether the join $(4K_2) + N_7$ is Hamiltonian. Justify your answer.*

Solution. (i) $(3K_2) + N_7$ is not Hamiltonian.

Let S be the set of vertices in $3K_2$. So $|S| = 6$. Observe that if we remove all vertices of S from the graph $(3K_2) + N_7$, we will obtain the graph N_7, which has 7 components. By Theorem 6.3, $(3K_2) + N_7$ is not Hamiltonian.

(ii) $(4K_2) + N_7$ is Hamiltonian. Note that the order of $(4K_2) + N_7$ is 15 and the minimum degree of this graph is 8 ($\geq 15/2$). By Theorem 6.4, $(4K_2) + N_7$ is Hamiltonian. (Indeed this graph is 8-regular.) $\square$

Chapter 7

Digraphs and Tournaments

Theorem 7.1 *Let T be a tournament. Then T is transitive if and only if T contains no cycles.*

Theorem 7.2 *Every tournament contains a Hamiltonian path.*

Theorem 7.3 *Let T be a tournament. If w is a vertex in T with maximum out-degree, then w is a king in T.*

Result (1). Let D be a digraph. Then

$$\sum_{v \in V(D)} id(v) = e(D) = \sum_{v \in V(D)} od(v).$$

Result (2). Let T_n denote a tournament of order n. Then

(i) $e(T_n) = \binom{n}{2}$;

(ii) $od(v) + id(v) = n - 1$ for each vertex v in T_n; and

(iii) $\displaystyle\sum_{v \in V(T_n)} od(v) = \binom{n}{2} = \sum_{v \in V(T_n)} id(v).$

Remark 7.1. Throughout this chapter, we shall assume the following:
All digraphs contain neither parallel arcs nor loops.

Exercise 7.1

Problem 1. *Let D be the digraph shown below:*

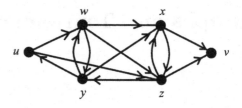

Find

(i) *V(D) and E(D).*

(ii) *v(D) and e(D).*

(iii) *all vertices adjacent from w.*

(iv) *all vertices adjacent to y.*

(v) *all vertices dominated by x.*

(vi) *all vertices that dominate z.*

(vii) *all arcs incident from u.*

(viii) *all arcs incident to z.*

Solution. (i)

$$V(D) = \{u, v, w, x, y, z\}$$

and

$$E(D) = \{uw, uz, wx, wy, wz, xv, xz, yu, yw, yx, zv, zx, zy\}.$$

(ii) $v(D) = 6$ and $e(D) = 13$.

(iii) x, y, z are the vertices adjacent from w.

(iv) z and w are the vertices adjacent to y.

(v) v and z are the vertices dominated by x.

(vi) u, w and x are the vertices that dominate z.

(vii) uw and uz are the arcs incident from u.

(viii) uz, wz and xz are the arcs incident to z.   □

Problem 2. *Let D be the digraph defined as follows:*

$$V(D) = \{a, b, u, v, w, x, y, z\}$$

and

$$E(D) = \{aw, ay, bx, ux, vz, wu, wx, wy, xz, ya, yv, yx, yz, zb, zy\}.$$

(i) *Draw the diagram of D.*

(ii) *Find $v(D)$ and $e(D)$.*

(iii) *Which vertices are adjacent from a?*

(iv) *Which vertices are adjacent to y?*

(v) *Which vertices are dominated by w?*

(vi) *Which vertices dominate x?*

(vii) *Which arcs are incident to z?*

(viii) *Which arcs are incident from y?*

Solution. (i)

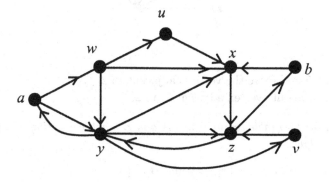

(ii) $v(D) = 8$ and $e(D) = 15$.

(iii) w and y are adjacent from a.

(iv) a, w and z are adjacent to y.

(v) u, x and y are dominated by w.

(vi) b, u, w and y dominate x.

(vii) xz, yz and vz are the arcs incident to z.

(viii) ya, yv, yx and yz are the arcs incident from y. ▫

Problem 3. *Let D be the digraph defined as follows:*

$$V(D) = \{1, 2, 3, 4, 5, 6\}$$

and $(i, j) \in E(D)$, where i, j are in $V(D)$, if and only if $i > j$.

  (i) *Draw the diagram of D.*

 (ii) *Find $e(D)$.*

(iii) *Which vertices are adjacent to '2'?*

(iv) *Which vertices are adjacent from '2'?*

Solution. (i)

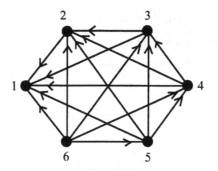

(ii) $e(D) = 15$.

(iii) $3, 4, 5$ and 6 are the vertices adjacent to 2.

(iv) 1 is the only vertex adjacent from 2.   □

Problem 4. *Two table tennis teams A and B, each consisting of 3 players as shown below:*

$$A = \{x, y, z\} \text{ and } B = \{u, v, w\},$$

had a friendly match between their players in singles. Each player in a team must play each player in the other exactly once with no ties allowed. At the end of the match, it was reported that

  (i) *x won all the matches;*

 (ii) *y was defeated only by w;*

(iii) *team A defeated team B by just one match.*

(1) Construct a digraph D to model the situation where $V(D)$ is the set of all players, and 'a' → 'b' in D if player 'a' defeated player 'b'.

(2) Find $e(D)$.

(3) Did 'z' win any game?

(4) Which player in team B won the largest number of matches?

Solution. (1)

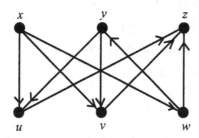

(2) $e(D) = 9$.

(3) z did not win any game.

(4) In team B, w won the largest number of matches. □

Exercise 7.2

Problem 1. *Let D be the digraph considered in Problem 1 of Exercise 7.1.*

(i) *Find the in-degree and out-degree of each vertex in D.*

(ii) *Verify that the equalities in Result (1) hold.*

(iii) *Is there any source in D?*

(iv) *Is there any sink in D?*

(v) *For $k = 2, 3, 4, 5$, find a k-cycle in D.*

(vi) *Is there any 6-cycle in D? Why?*

(vii) *Which vertices are reachable from u?*

(viii) *Which vertices are reachable from v?*

(ix) *Is D connected?*

(x) *Is D strongly connected?*

(xi) *Find $d(u, x), d(u, z), d(u, v), d(x, w)$ and $d(v, x)$.*

Solution. (i) $id(u) = 1$ and $od(u) = 2$,
   $id(v) = 2$ and $od(v) = 0$,
   $id(w) = 2$ and $od(w) = 3$,
   $id(x) = 3$ and $od(x) = 2$,
   $id(y) = 2$ and $od(y) = 3$,
   $id(z) = 3$ and $od(z) = 3$.
(ii) Result (1) holds as $\sum\limits_{a \in V(D)} id(a) = \sum\limits_{a \in V(D)} od(a) = e(D) = 13$.
(iii) There is no any source in D, since $id(a) > 0$ for every $a \in V(D)$.
(iv) There is only one sink, i.e., v, as $od(v) = 0$.
(v) In D, ywy is a 2-cycle, $yuwy$ is a 3-cycle, $ywxzy$ is a 4-cycle and $yuwxzy$ is a 5-cycle.
(vi) There is no 6-cycle in D. Since $od(v) = 0$, no cycle can include v, implying that the longest cycle in D contains at most 5 vertices.
(vii) All vertices in D are reachable from 'u'.
(viii) Only v itself is reachable from 'v'.
(ix) Yes, D is connected.
(x) No, D is not strongly connected.
(xi) $d(u, x) = 2, d(u, z) = 1, d(u, v) = 2, d(x, w) = 3$ and $d(v, x) = \infty$. $\square$

Problem 2. *Let D be the digraph considered in Problem 2 of Exercise 7.1.*

(i) *Find the in-degree and out-degree of each vertex in D.*

(ii) *Verify that the equalities in Result (1) hold.*

(iii) *Is there any source in D?*

(iv) *Is there any sink in D?*

(v) *Find a 6-cycle in D.*

(vi) *Is there any 7-cycle in D?*

(vii) *Is there any 8-cycle in D?*

(viii) *Which vertices are reachable from u?*

(ix) *Is D connected?*

(x) *Is D strongly connected?*

(xi) *Find $d(a, u), d(u, a), d(a, b)$ and $d(b, a)$.*

(xii) *Find two vertices in D such that the distance from one of them to the other is 5.*

(xiii) *Find two vertices in D such that the distance from one of them to the other is 6.*

Solution. (i) $id(a) = 1; od(a) = 2;$
$id(b) = 1; od(b) = 1;$
$id(u) = 1; od(u) = 1;$
$id(v) = 1; od(v) = 1;$
$id(w) = 1; od(w) = 3;$
$id(x) = 4; od(x) = 1;$
$id(y) = 3; od(y) = 4;$
$id(z) = 3; od(z) = 2.$

(ii) Result (1) holds, as $\sum_{t \in V(D)} id(t) = \sum_{t \in V(D)} od(t) = e(D) = 15.$

(iii) There is no source in D, since $id(t) > 0$ for each $t \in V(D)$.

(iv) There is no sink in D, as $od(t) > 0$ for each $t \in V(D)$.

(v) $awuxzya$ is a 6-cycle in D.

(vi) There is no 7-cycle in D.

(vii) There is no 8-cycle in D.

(viii) all vertices in D are reachable from 'u'.

(ix) Yes, D is connected.

(x) Yes, D is strongly connected.

(xi) $d(a, u) = 2, d(u, a) = 4, d(a, b) = 3$ and $d(b, a) = 4$.

(xii) Take u, w and we have $d(u, w) = 5$.

(xiii) Take b, u and we have $d(b, u) = 6$.

Problem 3. *Let D be the digraph considered in Problem 3 of Exercise 7.1.*

(i) *Find $e(D)$.*

(ii) *Find the in-degree and out-degree of each vertex in D.*

(iii) *Verify that the equalities in Result (1) hold.*

(iv) *Is there any source in D?*

(v) *Is there any sink in D?*

(vi) *Are there any two vertices in D which have the same out-degree?*

(vii) *Is there any cycle in D?*

(viii) *Is there any path of length 5 in D?*

(ix) *What is $G(D)$?*

(x) *Is D strong?*

Solution. (i) $e(D) = 15$.

(ii) $id(1) = 5, od(1) = 0$,

$id(2) = 4, od(2) = 1$,

$id(3) = 3, od(3) = 2$,

$id(4) = 2, od(4) = 3$,

$id(5) = 1, od(1) = 4$,

$id(6) = 0, od(6) = 5$.

(iii) Result (1) holds as

$$\sum_{i \in V(D)} id(i) = \sum_{i \in V(D)} od(i) = e(D) = 15.$$

(iv) There is only one source in D, i.e., the vertex 6.

(v) There is only one sink in D, i.e., the vertex 1.

(vi) Every two vertices have different out-degrees.

(vii) There is no cycle in D.

(viii) Yes, there is a path of length 5, namely, 654321.

(ix) $G(D)$ is the complete graph K_6.

(x) D is not strong.   □

Problem 4. *Are the following two digraphs isomorphic? If 'YES', find an isomorphism from one of them to the other.*

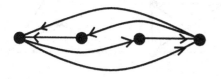
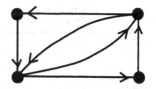

Solution. Yes, the two digraphs D and D' are isomorphic. First we name their vertices as shown below.

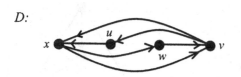
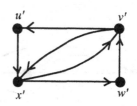

Define a mapping $f: V(D) \longrightarrow V(D')$ by $f(u) = u', f(v) = v', f(w) = w', f(x) = x'$. It can be shown that f is an isomorphism from D to D'. $\square$

Problem 5. *Are the following two digraphs isomorphic? Justify your answer.*

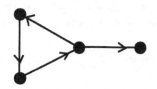
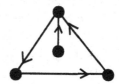

Solution. The two digraphs are not isomorphic, as one of them has a vertex with out-degree 0 while the other does not have any. $\square$

Problem 6. *Among the following three digraphs, which two are isomorphic? Justify your answer.*

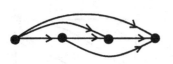 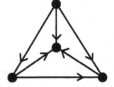

Solution. Both the first digraph and the third digraph have a vertex with in-degree 3, but the second digraph does not have. Thus the second digraph is not isomorphic to the first and the third.

Now we show that the first digraph and the third digraph are isomorphic. First name the vertices in the two digraphs as shown below.

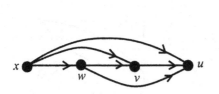 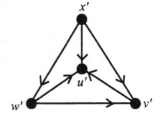

Then it can be shown that the mapping f defined by

$$f(u) = u', f(v) = v', f(w) = w', f(x) = x',$$

is an isomorphism from the first digraph to the third one. $\square$

Problem 7. (+) *Let D be a digraph. Prove that every $u - v$ walk in D contains a $u - v$ path, where u and v are two vertices in D.*

Solution. Let W_1 be a $u - v$ walk in D. If W_1 contains no cycles, then W_1 is a path in D.

Now assume that W_1 contains a cycle. Then some vertex in the cycle is repeated. Let x be such a vertex. Let W' be the $u - x$ walk along W_1 in which x is not repeated. Let W'' be the $x - v$ walk along W_1 in which x is not repeated. Then the walk $W_2 = W'W''$ is a $u - v$ walk whose arcs are in W_1. Note that W_2 is shorter than W_1.

This procedure is repeatedly applied until the resulting $u - v$ walk contains no cycle, and in this case, the resulting $u - v$ walk is a desired path. $\square$

Problem 8. (+) *Let D be a digraph and $k = \min\{od(v)|v \in V(D)\}$. Show that D contains a path of length at least k. Is the result also true if $k = \min\{id(v)|v \in V(D)\}$?*

Solution. Let $P = x_0 x_1 \cdots x_s$ be a longest path in D. As P is a longest path, x_s is not adjacent to any vertex in $V(D)\backslash\{x_0, x_2, \cdots, x_{s-1}\}$.

Thus, as $od(x_s) \geq k$, there exist at least k vertices in $\{x_0, x_1, \cdots, x_{s-1}\}$ which are adjacent from x_s. Hence $s \geq k$, implying that the length of P is at least k.

Similarly, it can be shown that if $k = \min\{id(v)|v \in V(D)\}$, then D contains a path of length at least k. □

Problem 9. (+) *Let D be a digraph and $k = \min\{od(v)|v \in V(D)\}$. Show that D contains an r-cycle, where $r \geq k + 1$. Is the result also true if $k = \min\{id(v)|v \in V(D)\}$?*

Solution. By the result in Problem 8, we know that D contains a longest path $P : x_0 x_1 \cdots x_s$ such that $s \geq k$.

As P is a longest path in D, all vertices adjacent from x_s are in the set $\{x_0, x_1, \cdots, x_{s-1}\}$.

Since $od(x_s) \geq k$, at least k vertices in $\{x_0, x_1, \cdots, x_{s-1}\}$ are adjacent from x_s. Thus there exists a vertex x_i with $0 \leq i \leq s - k$ such that $x_s x_i$ is an arc in D. Hence we get a cycle of length r:

$$x_i x_{i+1} \cdots x_s x_i,$$

where $r = s - i + 1 \geq k + 1$.

Likewise, if $k = \min\{id(v)|v \in V(D)\}$, we can show that the result also holds. □

Problem 10. *Let D be a digraph with $v(D) \geq 2$. Prove that if D is strong, then $id(v) \geq 1$ and $od(v) \geq 1$ for each vertex v in D (that is D contains neither sink nor source). Is the converse true?*

Solution. If D contains a sink x, i.e., $od(x) = 0$, then no vertex other than x is reachable from x and so D is not strong. Likewise, if D contains a source, then D is not strong.

Hence if D is strong, then $id(v) \geq 1$ and $od(v) \geq 1$ for each vertex v in D.

The converse is not true. The following digraph is an example.

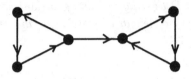

Problem 11. (+) *Let D be a digraph whose underlying graph G(D) is the following 2 × 3 grid:*

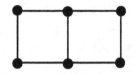

Assume that D contains neither sink nor source. Show that D contains a 4-cycle.

Solution. Suppose that D contains no 4-cycles. We first label the vertices in this digraph as shown below:

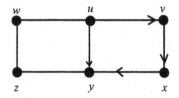

Without loss of generality, assume that $v \to x$. As D contains neither sink nor source, $u \to v$ and $x \to y$. By the assumption that D contains no 4-cycles, $u \to y$, as shown in the diagram.

As D contains no sink, $y \to z$, and in turn, $z \to w$ and $w \to u$. But then $uyzwu$ is a 4-cycle in D, a contradiction.

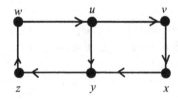

Problem 12. (+) *Let D be a digraph whose underlying graph G(D) is K(4, 4). Prove that if D contains an 8-cycle, then D contains a 4-cycle.*

Solution. Assume that D contains an 8-cycle, but D contains no 4-cycles. We may assume that $uu'vv'ww'xx'u$ is an 8-cycle as shown below:

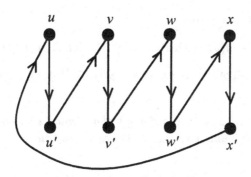

Since D contains no 4-cycles, $u \to v'$; otherwise, $uu'vv'u$ is a 4-cycle. By the same reason, $v \to w'$, $w \to x'$ and $x \to u'$, as shown below:

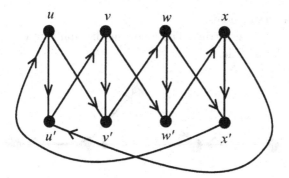

However, we find a 4-cycle, namely, $uv'wx'u$, a contradiction. □

Problem 13. (+) *Let D be a digraph of order $n \geq 2$. Assume that $od(x) + id(y) \geq n - 1$ for any two vertices x, y in D such that x is not adjacent to y. Prove that D is strong.*

Solution. We claim that for any two vertices u, v in D, if u is not adjacent to v, then uwv is a path in D for some w in D.

Let $O(u)$ be the set of vertices in D which are adjacent from u and $I(v)$ be the set of vertices in D which are adjacent to v. As u is not adjacent to v, $v \notin O(u)$ and $u \notin I(v)$. So

$$O(u) \cup I(v) \subseteq V(D)\backslash\{u, v\}.$$

If $O(u) \cap I(v) = \emptyset$, then

$$od(u) + id(v) = |O(u)| + |I(v)| \leq |V(D)\backslash\{u, v\}| = n - 2,$$

a contradiction. Thus there exists $w \in O(u) \cap I(v)$, implying that uwv is a path in D. $\square$

Problem 14. (+) *Let D be a digraph that contains no cycles. Prove that D contains a sink and a source. Is the converse true?*

Solution. Suppose on the contrary that D contains no sink; i.e., $od(v) > 0$ for every vertex v in D.

Let $k = \min\{od(v)|v \in V(D)\}$. Then $k \geq 1$. By the result in Problem 9, D contains an r-cycle, where $r \geq k + 1$, a contradiction.

Similarly, if $id(v) > 0$ for every vertex v in D, then D also contains a cycle.

Hence the result holds.

The converse is not true. The following digraph contains a source and a sink, and also a cycle.

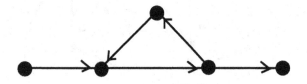

$\square$

Problem 15. (+) *Let D be a digraph. Prove that D contains no cycles if and only if every walk in D is a path.*

Solution. ($\Leftarrow$) If D contains a cycle, say $v_1 v_2 \cdots v_k v_1$, then $v_1 v_2 \cdots v_k v_1 v_2$ is a $v_1 - v_2$ walk which is not a path.

($\Rightarrow$) Let W be a walk in D that is not a path. Then some vertex is repeated in W. Let v be a vertex in W which is repeated. We may choose v so that no vertex in the section (*) (see the diagram below) is repeated in (*).

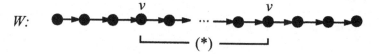

Clearly, the closed walk of (*) forms a cycle in D. $\qquad\square$

Problem 16. (+) *Construct a digraph D of order 7 such that $id(v) = od(v) = 2$ for each vertex v in D, but D contains no k-cycles, where $k = 2, 4, 6$.*

Solution. The following digraph is such a digraph.

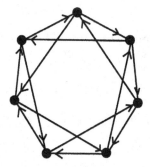

$\qquad\square$

Problem 17. *Let D be a digraph and A, a set of vertices in D. Denote by $R(A)$ the set of vertices in D which are reachable from some vertex in A. Clearly, $A \subseteq R(A)$, as every vertex is reachable from itself.*
(a) Consider the following digraph D:

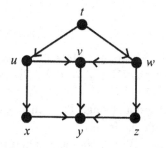

(i) *Let $S = \{x, z\}$, $T = \{u, w\}$ and $U = \{x, y, z\}$. Find $R(S)$, $R(T)$ and $R(U)$.*

(ii) *Is $R(U) = U$?*

(iii) *Is D strong?*

(b) *Consider the following digraph D:*

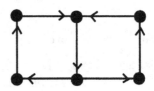

(i) *Can you find a non-empty set W of vertices in D such that $W \neq V(D)$ and $R(W) = W$?*

(ii) *Is D strong?*

(c) (+) *Prove that a digraph D is strong if and only if for every non-empty and proper subset W of $V(D)$, $W \subset R(W)$.*

Solution.   (a)

(i) $R(S) = \{x, z, y\}$, $R(T) = \{u, v, w, x, y, z\}$, $R(U) = \{x, y, z\}$.

(ii) Yes, $R(U) = U$.

(iii) D is not strong.

(b)

(i) No, there is no non-empty subset W of vertices in D such that $W \neq V(D)$ and $R(W) = W$.

(ii) Yes, D is strong.

(c)

($\Rightarrow$) Assume that D is strong. Then any two vertices are mutually reachable in D. Thus for any W, $\emptyset \neq W \subset V(D)$, we have $R(W) = V(D)$. Hence the necessity holds.

($\Leftarrow$) Suppose that D is not strong. Then there exist two vertices u, v in D such that u is not reachable from v.

Let $W = R(\{v\})$. Then W is a non-empty and proper subset of $V(D)$.

We claim that $R(W) = W$. Let $x \in R(W)$. Then x is reachable from a vertex y in W. As $y \in W = R(\{v\})$, y is reachable from v. Hence x is reachable from v, i.e., $x \in R(\{v\}) = W$. This shows that $W = R(W)$.

So the sufficiency holds. □

Problem 18. *An orientation of a graph G is a digraph D obtained from G by assigning each edge in G an arbitrary direction. (Clearly, the underlying graph of D is G.) For instance, a graph G and an orientation of G are shown below.*

 G:  D: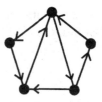

(i) *Let G be a graph. Does there exist an orientation D of G such that $d(x,y) \le 1$ for all x, y in $V(D)$?*

(ii) *For $n = 3, 5$ and 6, find an orientation D_n of K_n such that $d(x,y) \le 2$ for all x, y in $V(D_n)$.*

(iii) *Find an orientation D_4 of K_4 such that $d(x,y) \le 3$ for all x, y in $V(D_4)$.*

(iv) *(+) Does there exist an orientation D_4 of K_4 such that $d(x,y) \le 2$ for all x, y in $V(D)$?*

(v) *(+) Suppose that, for some $n \ge 5$, K_n has an orientation D_n such that $d(x,y) \le 2$ for all x, y in $V(D_n)$. Construct an orientation D_{n+2} of K_{n+2} based on D_n such that $d(x,y) \le 2$ for all x, y in $V(D_{n+2})$.*

Solution. (i) For any graph G of order at least 2, there is no orientation D such that $d(x,y) \le 1$ for all x, y in $V(D)$.

The reason is very simple. If $d(x,y) = 1$, then it is impossible that $d(y,x) = 1$.

(ii) An orientation D_n of K_n, $n = 3, 5, 6$, such that $d(x,y) \le 2$ for any $x, y \in V(D_n)$, is given below:

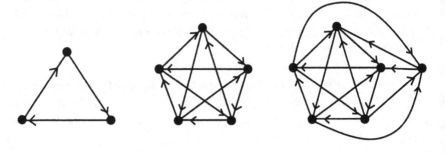

(iii) An orientation D_4 of K_4 such that $d(x,y) \leq 3$ for any $x,y \in V(D_4)$ is given below:

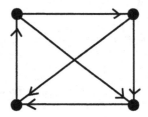

(iv) No, there is no orientation D_4 of K_4 such that $d(x,y) \leq 2$ for any $x,y \in V(D_4)$.

Suppose on the contrary that such a D_4 exists. Let yx be an arc in D_4. Then $d(x,y) = 2$, and so $xzyx$ is a 3-cycle in D_4 for some vertex z as shown in (a).

Let u be the fourth vertex. Then u must be adjacent to some vertex in $\{x,y,z\}$, say x (by symmetry). Then $d(x,u) = 2$, implying that $z \rightarrow u$, as shown in (b).

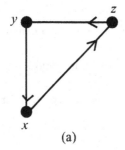

(a)

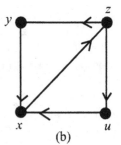

(b)

Now consider y and u. If $y \rightarrow u$, then $d(u, y) = 3$; if $u \rightarrow y$, then $d(y, u) = 3$, a contradiction.

(v) An orientation D_{n+2} of K_{n+2} is constructed from D_n by adding two new vertices u and v and adding all arcs in the set $\{uv\} \cup \{vw, wu | w \in V(D_n)\}$, as shown below.

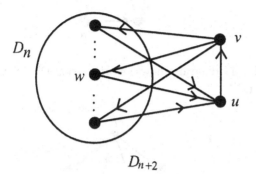

D_{n+2}

Now we show that $d(x, y) \leq 2$ for all $x, y \in V(D_{n+2})$. This is obvious if $\{x, y\} \cap \{u, v\} = \emptyset$.

By the definition of D_{n+2}, $uvwu$ is a 3-cycle of D_{n+2} for every $w \in V(D_n)$. Thus $d(u, v) = 1$, $d(v, u) = 2$ and

$$d(u, w) = 2, d(w, u) = 1, d(v, w) = 1, d(w, v) = 2,$$

for every $w \in V(D_n)$. Hence $d(x, y) \leq 2$ for all $x, y \in V(D_{n+2})$ if $\{x, y\} \cap \{u, v\} \neq \emptyset$. $\qquad\square$

Problem 19. *Consider the orientation D of the graph G shown at the beginning of Problem 18.*

(i) *Verify that $d(x, y) \leq 4$ in D for all x, y in $V(D)$.*

(ii) *Is it true that $d(x, y) \leq 3$ in D for all x, y in $V(D)$?*

(iii) *Find an orientation D' of G such that $d(x, y) \leq 3$ in D' for all x, y in $V(D')$.*

(iv) *Does there exist an orientation D^* of G such that $d(x, y) \leq 2$ in D^* for all x, y in $V(D^*)$? Justify your answer.*

Solution. (i) As D is strongly connected and D contains only 5 vertices, $d(x, y) \leq 4$ for all x, y in $V(D)$.

(ii) No, we have $d(a, c) = 4$ (see the following diagram).

D:

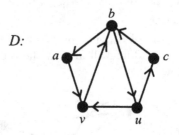

(iii) The following diagram shows an orientation D' such that $d(x, y) \leq 3$ for all x, y in $V(D')$.

D':

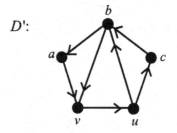

(iv) No, the graph G has no orientation D such that $d(x, y) \leq 2$ for all x, y in $V(D)$.

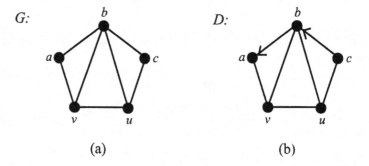

(a)   (b)

Suppose on the contrary that such a D exists. Consider a and c. As $d(c, a) \leq 2$, we must have $c \to b \to a$ (see the above diagram) in D. But then $d(a, c) \leq 2$ can never be the case, a contradiction.   □

Problem 20. *Let $G = K(p,p)$, where $p \geq 2$.*

(i) *Does there exist an orientation D of G such that $d(x,y) \leq 2$ in D for all x, y in $V(D)$? Justify your answer.*

(ii) *Find an orientation D of G such that $d(x,y) \leq 3$ in D for all x,y in $V(D)$.*

Solution. (i) There is no orientation D of G such that $d(x,y) \leq 2$ for all x, y in $V(D)$.

Let x, y be two vertices in D with $x \rightarrow y$. Then $d(y,x) \geq 2$. If $d(y,x) = 2$, then there exists a 3-cycle $xywx$ in D, contradicting the fact that G is bipartite.

(ii) Let (X, Y) be the bipartition of $K(p,p)$, where $X = \{a_1, a_2, \cdots, a_p\}$ and $Y = \{b_1, b_2, \cdots, b_p\}$. Let D be the orientation with arc set:

$$\{a_i b_i \mid i = 1, 2, \cdots, p\} \cup \{b_i a_j \mid 1 \leq i, j \leq p, i \neq j\}.$$

For any i, j with $i \neq j$, $a_i b_i a_j b_j a_i$ is a 4-cycle. Thus $d(a_i, a_j) \leq 3$, $d(b_i, b_j) \leq 3$, $d(a_i, b_j) \leq 3$ and $d(b_i, a_j) \leq 3$. Hence $d(x,y) \leq 3$ in D for all x,y in $V(D)$. □

Problem 21. *Let D be a digraph with n vertices labeled $v_1, v_2, \cdots, v_n$. The* **adjacency matrix** *of D, denoted by $A(D)$, is the $n \times n$ matrix in which $a_{i,j}$, the entry in row i and column j, is 1 if there is an arc from vertex v_i to vertex v_j, and 0 otherwise. We may sometimes write $A(D) = (a_{i,j})$.*

(i) *Draw the digraph D which has the following adjacency matrix $A(D)$:*

$$A(D) = \begin{pmatrix} 0 & 1 & 0 & 1 \\ 0 & 0 & 1 & 1 \\ 0 & 1 & 0 & 0 \\ 0 & 1 & 1 & 0 \end{pmatrix}$$

(ii) *Find the adjacency matrix $A(D)$ of the following digraph D.*

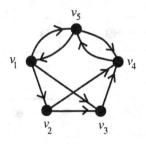

Solution. (i) The digraph is shown below:

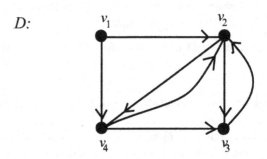

$$D:$$

(ii) The adjacency matrix of D is $A(D) = \begin{pmatrix} 0 & 1 & 1 & 0 & 1 \\ 0 & 0 & 1 & 1 & 0 \\ 0 & 0 & 0 & 1 & 0 \\ 0 & 0 & 0 & 0 & 1 \\ 1 & 0 & 0 & 1 & 0 \end{pmatrix}.$ □

Problem 22. *Let D be a digraph of order n, where $n \geq 2$.*

(i) *Show that if D contains no cycles, then $od(v) = 0$ for some vertex v in D.*

(ii) *(+) Show that D contains no cycles if and only if the vertices in D can be named as $v_1, v_2, \cdots, v_n$ such that $A(D)$ is upper triangular.*
*(Note that a square matrix $(a_{i,j})$ is called an **upper triangular matrix** if $a_{i,j} = 0$ for all i, j with $i > j$.)*

Solution. (i) If $od(v) \geq 1$ for all $v \in V(D)$, then, by the result of Problem 9, D contains a cycle of length at least 2, a contradiction.

(ii) ($\Rightarrow$) Assume that D contains no cycles.

We shall prove the result by induction on $n \geq 2$. For $n = 2$, D is either

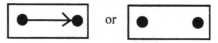

or

Name the vertices of D as

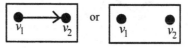

or

Then

$$A(D) = \begin{pmatrix} 0 & 1 \\ 0 & 0 \end{pmatrix} \text{ or } A(D) = \begin{pmatrix} 0 & 0 \\ 0 & 0 \end{pmatrix}.$$

Now assume that $n \geq 3$. By the result in (i), there is a vertex v such that $od(v) = 0$. Label this vertex as v_n.

Let D' be the digraph obtained from D by deleting v_n. It is clear that D' contains no cycles. Thus, by the induction hypothesis, the vertices in D' can be named as $v_1, v_2, \cdots, v_{n-1}$ such that $A(D')$ is upper triangular. Hence $A(D)$ is also upper triangular, as

$$A(D) = \left(\begin{array}{c|c} A(D') & \begin{matrix} a_{1,n} \\ \vdots \\ a_{n-1,n} \end{matrix} \\ \hline 0 \cdots 0 & 0 \end{array} \right).$$

($\Leftarrow$) Assume that the vertices in D can be named as $v_1, v_2, \cdots, v_n$ such that $A(D)$ is upper triangular. Then the (i, j)-entry in $A(D)$ is 0 if $i \geq j$, implying that $v_i v_j$ is not an arc in D whenever $i \geq j$. If D contains a cycle $v_{i_1} v_{i_2} \cdots v_{i_k} v_{i_1}$, then $i_1 < i_2 < \cdots < i_k < i_1$, which is impossible. $\square$

Exercise 7.3

Problem 1. *Let T be a tournament, and u, v be two mutually reachable vertices in T. Prove that*

(i) *$d(u, v) \neq 1$ if and only if $d(v, u) = 1$;*

(ii) *$d(u, v) \neq d(v, u)$.*

Solution. (i) If $d(u, v) \neq 1$, then u is not adjacent to v in T. As T is a tournament, $v \to u$ in T, and so $d(v, u) = 1$.

If $d(v, u) = 1$, then $v \to u$ in T, and so u is not adjacent to v in T, implying that $d(u, v) \neq 1$.

(ii) It follows from (i). $\square$

Problem 2. (+) *Let T be a tournament, and u, v be two vertices in T with $d(u, v) = k \geq 2$. Prove that*

(i) *$od(v) \geq k - 1$;*

(ii) *v is contained in an r-cycle for each $r = 3, 4, \cdots, k + 1$;*

(iii) *u and v are contained in a common $(k + 1)$-cycle.*

Solution. As $d(u, v) = k \geq 2$, let $u_1 \cdots u_k v$ be a shortest $u - v$ path in T, where $u_1 = u$. Then $v \to u_i$ for each $i = 1, 2, \cdots, k - 1$. Thus, we have
(i) $od(v) \geq k - 1$;
(ii) for $r = 3, 4, \cdots, k + 1$, $v u_{k-r+2} u_{k-r+3} \cdots u_k v$ is a r-cycle that contains v;
(iii) $v u_1 u_2 \cdots u_k v$ is a $(k + 1)$-cycle that contains both u and v. □

Problem 3. *Show that, up to isomorphism,*

(i) *there is only one strong tournament of order 3;*

(ii) *there is only one strong tournament of order 4.*

Solution. (i) Let T_3 be a strong tournement of order 3 and x_1, x_2, x_3 be the vertices in T_3.

As T_3 is strong, $od(x_i) = id(x_i) = 1$ for $i = 1, 2, 3$. Assume that $x_1 \to x_2$. Then $x_2 \to x_3$ and $x_3 \to x_1$. Thus T_3 is uniquely determined by the condition that $x_1 \to x_2$, as shown in (a). This shows that, up to isomorphism, there is only one strong tournament of order 3.

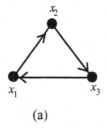

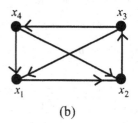

(a)   (b)

(ii) Let T_4 be a strong tournament of order 4 and x_1, x_2, x_3, x_4 be the vertices in T_4.

By the result in Problem 18 (iv) of Exercise 7.2, $d(x_i, x_j) = 3$ for some i, j, say $i = 1$ and $j = 4$. Let $x_1 x_2 x_3 x_4$ be a $x_1 - x_4$ path in T_4. As

$d(x_1, x_4) = 3$, we must have $x_4 \to x_1$, $x_4 \to x_2$ and $x_3 \to x_1$, as shown in (b).

Thus, up to isomorphism, there is only one strong tournament of order 4. □

Problem 4. *Let T_n be a strong tournament of order n such that, for each arc in T_n, the reversing of the direction of this arc results in also a strong tournament. Show that $n \geq 5$ and construct one such T_n.*

Solution. For $n = 3, 4$, by the result in the previous problem, T_3 and T_4 are the tournaments shown in the solution of the problem. Clearly, the reversing of $x_2 \to x_3$ in both T_3 and T_4 results in non-strong tournaments.

Hence $n \geq 5$. For $n = 5$, one such tournament is shown in (a).

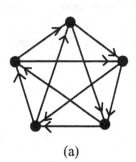
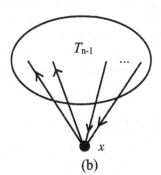

(a)   (b)

For $n \geq 6$, construct T_n from T_{n-1}, as shown in (b), by
(1) adding a new vertex x;
(2) adding two arcs from x to any two vertices in T_{n-1} and adding $n-3$ arcs from the other $n-3$ vertices of T_{n-1} to x.

It is easily checked that if T_{n-1} satisfies the condition stated in the problem, then so does T_n. □

Problem 5.

(i) *Suppose that six teams play in a round-robin tournament. Is it possible that all six teams have the same score at the end?*

(ii) *Suppose that seven teams play in a round-robin tournament. Is it possible that all seven teams have the same score at the end?*

(iii) *Suppose that n teams, $n \geq 3$, play in a round-robin tournament. Is it possible that all n teams have the same score at the end?*

Solution. (i) It is impossible. The total score of the six teams in the tournament is 15 and 15 is not divisible by 6.

(ii) It is possible. We can construct a tournament T of order 7 such that $id(u) = od(u) = 3$ for every vertex u in T. Let

$$V(T) = \{x_1, x_2, \cdots, x_7\}.$$

For each $i = 1, 2, \cdots, 7$, assume that x_i is adjacent only to x_{i+1}, x_{i+2} and x_{i+3}, where $x_{7+k} = x_k$ for all $k = 1, 2, 3$.

It can easily be checked that $id(x_i) = od(x_i) = 3$ for each $i = 1, 2, \cdots, 7$.

(iii) Let T_n be a tournament of order $n \geq 3$. By Result (2) (iii), $\sum\limits_{v \in V(T_n)} od(v) = \frac{n(n-1)}{2}$. Thus, all vertices have the same score $(= \frac{n-1}{2})$ only if $\frac{n-1}{2}$ is an integer, i.e., n is odd.

For odd $n \geq 3$, such a tournament can be constructed inductively.

For $n = 3$, it is clear that the strong tournament of order 3 is such a T_3.

Assume that there is such a tournament T_{2k+1}. Let $x_1, x_2, \cdots, x_{2k+1}$ be the vertices in T_{2k+1}. Now construct a tournament T_{2k+3} from T_{2k+1} by adding two new vertices u and v and adding arcs in the following set:

$$\{uv\} \cup \{ux_i, x_iv \mid i = 1, 2, \cdots, k\} \cup \{x_{i+k}u, vx_{i+k} \mid i = 1, 2, \cdots, k+1\}.$$

It can be checked that all vertices in T_{2k+3} have the same score. $\square$

Problem 6. *Let T be a tournament of order $n \geq 3$. Assume that $od(v) = k$ for all vertices v in T.*

(i) *Find a relation between k and n.*

(ii) *Deduce that n must be odd.*

(iii) *Construct one such tournament of order 7.*

Remark. *Such a tournament is called a **regular tournament**.*

Solution. (i) By Result (2) (iii) and the assumption,

$$\binom{n}{2} = \sum_{v \in V(T)} od(v) = kn.$$

Thus $k = (n-1)/2$.

(ii) By (i), $n = 2k + 1$, which is odd.

(iii) See the construction shown in Problem 5 (ii). $\square$

Problem 7. *Consider the following tournament T. Find*

(i) *the out-degree and in-degree of each vertex in T;*

(ii) *the sum $od(u) + od(v) + od(w) + od(x) + od(y)$;*

(iii) *the sum $id(u) + id(v) + id(w) + id(x) + id((y)$;*

(iv) *the sum $(od(u))^2 + (od(v))^2 + (od(w))^2 + (od(x))^2 + (od(y))^2$;*

(v) *the sum $(id(u))^2 + (id(v))^2 + (id(w))^2 + (id(x))^2 + (id(y))^2$.*

Are the sums in (ii) and (iii) the same?
Are the sums in (iv) and (v) the same?

T:

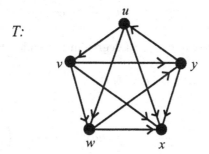

Solution. (i) In T, we have:

$$od(u) = 3, od(v) = 3, od(w) = 2, od(x) = 0, od(y) = 2;$$

$$id(u) = 1, id(v) = 1, id(w) = 2, id(x) = 4, id(y) = 2.$$

(ii)

$$od(u) + od(v) + od(w) + od(x) + od(y) = 10.$$

(iii)

$$id(u) + id(v) + id(w) + id(x) + id((y) = 10.$$

(iv)

$$(od(u))^2 + (od(v))^2 + (od(w))^2 + (od(x))^2 + (od(y))^2 = 3^2 + 3^2 + 2^2 + 0^2 + 2^2 = 26.$$

(v)

$$(id(u))^2 + (id(v))^2 + (id(w))^2 + (id(x))^2 + (id(y))^2 = 1^2 + 1^2 + 2^2 + 4^2 + 2^2 = 26.$$

The sums in (ii) and (iii) are the same, and the sums in (iv) and (v) are the same. $\square$

Problem 8. *Prove the results in Result (2).*

Solution. We shall prove that

(i) $e(T_n) = \binom{n}{2}$;

(ii) $od(v) + id(v) = n - 1$ for each vertex v in T_n; and

(iii) $\displaystyle\sum_{v \in V(T_n)} od(v) = \binom{n}{2} = \sum_{v \in V(T_n)} id(v)$.

(i) As the underlying graph of T_n is the complete graph K_n, we have $e(T_n) = e(K_n) = \binom{n}{2}$.

(ii) For each vertex v in T_n, its degree in the underlying graph of T_n is $n - 1$. Thus $od(v) + id(v) = n - 1$.

(iii) By Result (1) and Result (2) (i),

$$\sum_{v \in V(T_n)} od(v) = \sum_{v \in V(T_n)} id(v) = e(T_n) = \binom{n}{2}.$$

$\square$

Problem 9. (+) *Let T be a tournament. Show that*

$$\sum_{v \in V(T)} (od(v))^2 = \sum_{v \in V(T)} (id(v))^2.$$

(Putnam Exam (1965))

Solution. Let T be of order n. By Result (2), we have $od(v) + id(v) = n - 1$ for each vertex v in T and

$$\sum_{v \in V(T)} od(v) = \binom{n}{2} = \sum_{v \in V(T)} id(v).$$

Thus

$$\sum_{v \in V(T)} (od(v))^2 - \sum_{v \in V(T)} (id(v))^2$$

$$= \sum_{v \in V(T)} ((od(v))^2 - (id(v))^2)$$

$$= \sum_{v \in V(T)} (od(v) - id(v))(od(v) + id(v))$$

$$= (n-1) \sum_{v \in V(T)} (od(v) - id(v))$$

$$= (n-1) \left(\sum_{v \in V(T)} od(v) - \sum_{v \in V(T)} id(v) \right)$$

$$= (n-1) \left(\binom{n}{2} - \binom{n}{2} \right)$$

$$= 0. \qquad \square$$

Problem 10. (+) *Let T be a tournament with $V(T) = \{v_1, v_2, \cdots, v_n\}$. Show that, for any $k = 1, 2, \cdots, n$,*

$$\sum_{i=1}^{k} od(v_i) \geq \binom{k}{2}.$$

Solution. Let T_k be the sub-tournament of T induced by $\{v_1, v_2, \cdots, v_k\}$. Then, for $i = 1, 2, \cdots, k$,

$$od(v_i) \geq od_{T_k}(v_i),$$

where $od_{T_k}(v_i)$ is the out-degree v_i in T_k. Hence

$$\sum_{i=1}^{k} od(v_i) \geq \sum_{i=1}^{k} od_{T_k}(v_i) = \binom{k}{2}. \qquad \square$$

Problem 11. *Is there a tournament in which the out-degrees of the vertices are:*

(i) $5, 4, 3, 2, 1, 0$?

(ii) $5, 5, 3, 1, 1, 0$?

(iii) $5, 4, 4, 1, 1, 0$?

(iv) $4, 3, 3, 2, 2, 1$?

For each case, construct one such tournament if your answer is 'YES'; give a reason if your answer is 'NO'.

Solution. (i) There is a tournament in which the out-degrees of the vertices are $5, 4, 3, 2, 1, 0$.

Let T be the tournament with vertex set $\{x_1, x_2, \cdots, x_6\}$ and arc set

$$\{x_i x_j \mid 1 \le i < j \le 6\}.$$

Then $od(x_i) = 6 - i$ for $i = 1, 2, \cdots, 6$.

(ii) There is no tournament in which the out-degrees of the vertices are $5, 5, 3, 1, 1, 0$.

Suppose that T is such a tournament. There is a vertex x in T with $od(x) = 5$. Thus $x \to v$ for every other vertex v in T. But then $od(v) \le 4$ for each v with $v \ne x$, and so the second '5' can never be the out-degree of any vertex in T, a contradiction.

(iii) There is no tournament in which the out-degrees of the vertices are $5, 4, 4, 1, 1, 0$.

Suppose that T is such a tournament. Let x, y, z be the three vertices in T with out-degrees $1, 1, 0$, respectively. Then

$$od(x) + od(y) + od(z) = 1 + 1 + 0 = 2.$$

However, by the result of Problem 10, $od(x) + od(y) + od(z) \ge \binom{3}{2} = 3$, a contradiction.

(iv) There is a tournament T in which the out-degrees of the vertices are $4, 3, 3, 2, 2, 1$.

Let T be the tournament shown below:

T:

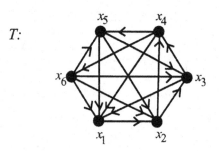

Then the out-degrees of $x_6, x_5, x_4, x_3, x_2, x_1$ are $4, 3, 3, 2, 2, 1$ respectively.

□

Problem 12.

(i) *Construct a tournament T_5 in which the out-degrees of the vertices are:* $2, 2, 2, 2, 2$.

(ii) *Construct a tournament T_6 in which the out-degrees of the vertices are:* $3, 3, 3, 2, 2, 2$.

(iii) *Show a way of combining the above T_5 and T_6 to obtain a tournament of order 11 in which the out-degrees of the vertices are:* $8, 8, 8, 8, 8,$ $3, 3, 3, 2, 2, 2$.

Solution.   (i) Let T_5 be the tournament with vertex set $\{x_1, x_2, x_3, x_4, x_5\}$ and arc set

$$E(T_5) = \{x_i x_j \mid i+1 \le j \le i+2, i = 1, 2, 3, 4, 5\},$$

where we assume that $x_6 = x_1$ and $x_7 = x_2$. Then $od(x_i) = id(x_i) = 2$ for all $i = 1, 2, 3, 4, 5$.

(ii) Let T_6 be the tournament with vertex set $\{y_1, y_2, y_3, y_4, y_5, y_6\}$ and arc set

$$E(T_6) = \{y_i y_j \mid i+1 \le j \le i+2, i = 1, 2, 3, 4, 5, 6\} \cup \{y_i y_{i+3} \mid i = 1, 2, 3\},$$

where we assume that $y_7 = y_1$ and $y_8 = y_2$.

Observe that $od(y_i) = 3$ for $i = 1, 2, 3$ and $od(y_i) = 2$ for $i = 4, 5, 6$.

(iii) Let T_{11} be the tournament with vertex set $V(T_{11}) = V(T_5) \cup V(T_6)$ and arc set

$$E(T_{11}) = E(T_5) \cup E(T_6) \cup \{x_i y_j \mid 1 \le i \le 5, 1 \le j \le 6\}.$$

Note that the out-degrees of $x_1, x_2, x_3, x_4, x_5, y_1, y_2, y_3, y_4, y_5, y_6$ in T_{11} are

$$8, 8, 8, 8, 8, 3, 3, 3, 2, 2, 2,$$

respectively.   □

Problem 13. (+) *Let T be a tournament of order $n \ge 3$. Prove that T contains a 3-cycle if and only if T contains two vertices of the same out-degree.*

Solution.   ($\Rightarrow$) Suppose that T does not contain two vertices of the same out-degree. Then the out-degrees of vertices in T are

$$0, 1, 2, \cdots, n-1.$$

Let $V(T) = \{x_1, x_2, \cdots, x_n\}$ and $od(x_i) = i - 1$. It can be shown by induction that

$$E(T) = \{x_j x_i \mid 1 \leq i < j \leq n\}.$$

Thus T contains no cycles, a contradiction.

($\Leftarrow$) Assume that T contains two vertices of the same out-degree.

Let x and y be two vertices in T with $od(x) = od(y) = k$. As either $x \to y$ or $y \to x$, $k \geq 1$. Assume that $x \to y$ and y is adjacent to k vertices $y_1, y_2, \cdots, y_k$. Since $od(x) = od(y) = k$ and $x \to y$, it is impossible that x is adjacent to every vertex in $\{y_1, y_2, \cdots, y_k\}$. Assume that $x \not\to y_1$. Then $y_1 \to x$, and so $x y y_1 x$ is a 3-cycle in T. □

Problem 14. (+) *Prove that a tournament T is transitive if and only if $od(u) \neq od(v)$ for any two vertices u, v in T.*

Solution. ($\Rightarrow$) Assume that T is transitive, i.e., for any three vertices u, v, w, if $u \to v$ and $v \to w$, then $u \to w$. Thus T contains no 3-cycles. By the result of Problem 13, $od(u) \neq od(v)$ for any two vertices u, v in T.

($\Leftarrow$) Assume that $od(u) \neq od(v)$ for any two vertices u, v in T. By the result of Problem 13, T contains no 3-cycles. Thus, for any three vertices u, v, w in T, if $u \to v$ and $v \to w$, we must have $u \to w$; that is, T is transitive. □

Problem 15. (+) *Prove that a tournament T of order n is transitive if and only if the out-degrees of its vertices are, respectively, $n-1, n-2, \cdots, 1, 0$.*

Solution. By the result of Problem 14, T is transitive if and only if $od(u) \neq od(v)$ for every two vertices u, v in T. Since T is of order n and $0 \leq od(u) \leq n-1$ for all $u \in V(T)$, $od(u) \neq od(v)$ for every two vertices u, v in T if and only if the out-degrees of vertices in T are, respectively, $n-1, n-2, \cdots, 1, 0$. □

Problem 16. (+) *A tournament T is said to be **reducible** if $V(T)$ can be partitioned into two non-empty subsets, U and W, such that $u \to w$ for all $u \in U$ and $w \in W$. A reducible tournament of order 5 is shown below:*

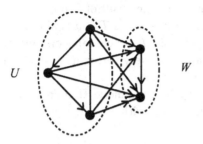

(i) Is the following tournament reducible?

(ii) Prove that every transitive tournament is reducible.

Solution. Notice that if T has a vertex of in-degree 0 or out-degree 0, then T is reducible.

(i) This tournament T is reducible, as it contains a vertex of in-degree 0.

(ii) Let T be a transitive tournament of order n. By the result of Problem 15, the out-degrees of vertices in T are $n-1, n-2, \cdots, 1, 0$. As T contains a vertex of out-degree 0, T is reducible. $\square$

Exercise 7.4

Problem 1. (+) *Prove that a tournament is transitive if and only if it has one and only one Hamiltonian path.*

Solution. ($\Rightarrow$) Let T be a tournament of order n. The proof is by induction on n. Assume that T is transitive. If $n \le 2$, then it is obvious that T contains one and only one Hamiltonian path.

Now assume that $n \ge 3$.

Since T is transitive, by the result of Problem 15 in Exercise 7.3, T has a vertex v with $od(v) = n - 1$. As $T - v$ is also a transitive tournament, by the induction hypothesis, $T - v$ contains one and only one Hamiltonian path. Since v is adjacent to every vertex in $T - v$, T also conatins one and only one Hamiltonian path.

($\Leftarrow$) Let $x_1 x_2 \cdots x_n$ be the unique Hamiltonian path in T. We shall show that $x_i \to x_j$ for all i, j with $1 \leq i < j \leq n$.

Clearly, it holds if $n \leq 2$.

Assume that $x_i \to x_j$ for all i, j with $1 \leq i < j \leq n - 1$. We need to show that $x_i \to x_n$ for all i with $1 \leq i \leq n - 1$.

Suppose on the contrary that $x_n \to x_i$ for some i with $1 \leq i \leq n - 1$. Let k be the minimum integer with $1 \leq k \leq n - 1$ such that $x_n \to x_k$.

If $k = 1$, then $x_n x_1 x_2 \cdots x_{n-1}$ is also a Hamiltonian path in T, a contradiction.

If $k \geq 2$, then $x_1 x_2 \cdots x_{k-1} x_n x_k x_{k+1} \cdots x_{n-1}$ is a Hamiltonian path in T, also a contradiction.

Hence $x_i \to x_n$ for all i with $1 \leq i \leq n - 1$.

Therefore $x_i \to x_j$ for all i, j with $1 \leq i < j \leq n$. This shows that T is transitive. $\qquad\square$

Problem 2. (+) *Let T be a tournament and u, v be two vertices in T. If $od(u) \geq od(v)$, what are the possible values of $d(u, v)$? Justify your answer.*

Solution. We claim that if $od(u) \geq od(v)$, then $d(u, v) \leq 2$.

If $u \to v$, then $d(u, v) = 1$. Assume that $v \to u$. We shall show that $d(u, v) = 2$.

Let $od(u) = k$ and assume that $u \to u_i$ for $i = 1, 2, \cdots, k$. As $od(v) \leq k$ and $v \to u$, it is impossible that $v \to u_i$ for all $i = 1, 2, \cdots, k$. Thus $u_i \to v$ for some i, where $i = 1, 2, \cdots, k$, say $u_1 \to v$. Then $u u_1 v$ is a $u - v$ path, implying that $d(u, v) = 2$ in this case. $\qquad\square$

Problem 3. (+) *Suppose in a round-robin tournament, team A has the maximum score. Let p denote the number of teams defeated by A, and q the number of teams not defeated by A. Which of the following situations are possible?*

(i) $p > q$;

(ii) $p = q$;

(iii) $p < q$.

Solution. If there are even number of teams in the round-robin tournament, only (i) is possible; otherwise, both (i) and (ii) are possible. But (iii) is impossible.

Assume that there are n teams in the round-robin tournament. Then $p + q = n - 1$ and

$$p \geq \binom{n}{2}/n = \frac{n-1}{2}.$$

Thus $q \leq (n-1)/2$. Hence $p \geq q$.

If n is even, we have $p \geq \lceil (n-1)/2 \rceil \geq n/2 > q$. If n is odd, then it is possible that $p = q = (n-1)/2$ or $p > (n-1)/2 \geq q$. $\square$

Problem 4. *For each $n \geq 2$, construct a tournament of order n in which there is a king w with $od(w) = 1$.*

Solution. The case that $n = 2$ is trivial.

Assume that $n \geq 3$. Let T_{n-2} be any tournament of order $n - 2$. Construct a tournament T_n from T_{n-2} by adding two new vertices w and u such that $w \rightarrow u$ and for all x in T_{n-2}, $u \rightarrow x \rightarrow w$ (see the diagram below).

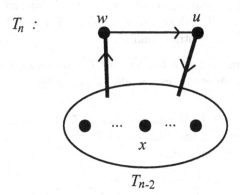

Clearly, w is a king and $od(w) = 1$ in T_n. $\square$

Problem 5. (+) *Let T be a tournament of order $n \geq 3$ and let u be a vertex in T with $od(u) \leq n - 2$. Show that u is dominated by a king in T.*

Solution. Assume that $od(u) = k$, where $0 \leq k \leq n - 2$.

Let $V(T) = \{u, x_1, x_2, \cdots, x_{n-1}\}$. Assume that $u \to x_i$ for $i = 1, 2, \cdots, k$, and $x_i \to u$ for $i = k+1, k+2, \cdots, n-1$.

Let T' be the sub-tournament of T induced by $\{x_{k+1}, x_{k+2}, \cdots, x_{n-1}\}$. Then T' itself has a king, say x_{n-1}.

So $d(x_{n-1}, x_i) \leq 2$ for all $i = k+1, k+2, \cdots, n-2$. Note also that $x_{n-1} \to u \to x_i$ for all $i = 1, 2, \cdots, k$. Thus x_{n-1} is a king of T.

This shows that u is dominated by a king of T.   □

Problem 6. *If w is the **only** king in a tournament of order n, what is the value of $od(w)$?*

Solution. If w is the **only** king in a tournament T of order n, then $od(w) = n - 1$. Otherwise, $od(w) \leq n - 2$, where $n \geq 3$, and by the result of Problem 5, T contains a king other than w.   □

Problem 7. (+) *Is there a tournament which contains exactly two kings? Justify your answer.*

Solution. No, no tournament contains exactly two kings.

Suppose on the contrary that T is a tournament of order $n \geq 2$ which contains exactly two kings, say u and v with $u \to v$. Since u is reachable from v, $od(u) \leq n-2$ and $n \geq 3$. By the result of Problem 5, u is dominated by a king, say w, in T. Clearly, $w \neq v$. Thus, T has at least three kings, namely, u, v and w, a contradiction.   □

Problem 8. *Suppose that a tournament T has exactly three kings. What can be said about the dominance relations among them?*

Solution. Let u, v, w be the three kings of T of order n. Let T^* be the sub-tournament of T induced by $\{u, v, w\}$. We claim that T^* is a 3-cycle. Otherwise, we may assume that

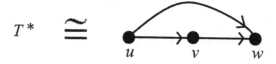

Since u is reachable from w in T, $od(u) \leq n-2$. By the result of Problem 5, u is dominated by a king, say z, in T. Clearly, $z \neq v$ and $z \neq w$. Thus, T contains more than three kings, a contradiction.   □

Problem 9. *Using the results in Problem 18 of Exercise 7.2, show that for each $n \geq 3$, $n \neq 4$, there is a tournament of order n in which every vertex is a king.*

Solution. By the results in Problem 18 of Exercise 7.2, if $n \geq 3$ and $n \neq 4$, there is a tournament T_n of order n such that $d(u, v) \leq 2$ for all $u, v \in V(T_n)$. Hence every vertex in T_n is a king. □

Problem 10. *Show that there is no tournament of order four in which every vertex is a king.*

Solution. This result follows from the result of Problem 18, Exercise 7.2.
    Another proof is given below.
    Assume that there is a tournament T of order four in which every vertex is a king. Let x_1, x_2, x_3, x_4 be the vertices in T. As every vertex is a king, $od(x_i) \geq 1$ for all i. Since

$$od(x_1) + od(x_2) + od(x_3) + od(x_4) = 6,$$

$od(x_i) = 1$ for some i.
    Assume that $od(x_1) = 1$ and $x_1 \to x_2$. Then $x_i \to x_1$ for $i = 3, 4$. As $d(x_1, x_i) \leq 2$ for $i = 3, 4$, we have $x_2 \to x_i$ for $i = 3, 4$.
    If $x_3 \to x_4$, then $d(x_4, x_3) = 3$; if $x_4 \to x_3$, then $d(x_3, x_4) = 3$. Both cases contradict the condition that every vertex is a king.
    Hence there is no such tournament T. □

Problem 11. *Let T be a regular tournament (see Problem 6 in Exercise 7.3). Is it true that every vertex in T is a king? Why?*

Solution. Yes, every vertex in a regular tournament T is a king.
    Since T is regular, by definition, there exists an integer k such that $od(v) = k$ for all vertices v in T. Thus, every vertex in T has the maximum out-degree, and so is a king by Theorem 7.3. □

Problem 12. (+) *A tournament T is said to be* **irreducible** *if T is* **not** *reducible, that is, for any partition of $V(T)$ into two non-empty subsets U and W, there exist $u \in U$, $w \in W$ such that $u \to w$ and there exist $y \in W$, $x \in U$ such that $y \to x$. (See Problem 16 in Exercise 7.3.)*

(i) *Is the following tournament irreducible?*

(ii) *Prove that a tournament is strong if and only if it is irreducible.*

Solution. (i) This tournament is irreducible.

(ii)

($\Rightarrow$) Assume that T is a strong tournament. Let (X, Y) be any partition of $V(T)$.

Let $x \in X$ and $y \in Y$. As T is strong, there exists a $x - y$ path in T. This path must contain an arc $x'y'$ with $y' \in Y$ and $x' \in X$.

Similarly, there exists an arc $y''x''$ with $y'' \in Y$ and $x'' \in X$.

Hence T is irreducible.

($\Leftarrow$) Assume that T is irreducible. To show that T is strong, we show that $R(x) = V(T)$ for all vertices x in T, where $R(x)$ is the set of all vertices in T which are reachable from x.

Suppose that $R(x)$ is a proper subset of $V(T)$ for some $x \in V(T)$. Since T is irreducible, there exists an arc $x'y'$ in T with $x' \in R(x)$ and $y' \in V(T)\backslash R(x)$. But this implies that $y' \in R(x)$, a contradiction. Hence $R(x) = V(T)$, and so T is a strong tournament. $\qquad\square$

Problem 13. *Let T be a tournament. Prove that if T is strong, then every vertex in T is contained in a cycle. Is the converse true?*

Solution. Let x be a vertex in T. As T is strong, $id(x) \geq 1$. Let y be a vertex such that $y \to x$.

As T is strong, there is a $x - y$ path P. The arc yx cannot be in P. Then P and the arc yx form a cycle containing x in T.

The following tournament shows that the converse is not true.

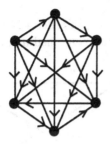

$\square$

Problem 14. (+) *Let T be a strong tournament of order $n \geq 3$. Determine whether each of the following statements is true.*

(i) *Every vertex in T is contained in a 3-cycle.*

(ii) *Every arc in T is contained in a 3-cycle.*

(iii) *Every arc in T is contained in a Hamiltonian cycle.*

(iv) *Every arc in T is contained in a cycle.*

(v) *For any two vertices u, v in T, either there is Hamiltonian path from u to v, or there is Hamiltonian path from v to u.*

Solution. (i) It is true. Let z be any vertex in T. Let

$$O(z) = \{x \in V(T) \mid z \to x\} \text{ and } I(z) = \{y \in V(T) \mid y \to z\}.$$

Since T is strong, both $O(z)$ and $I(z)$ are non-empty. Also, there exist $x \in O(z)$ and $y \in I(z)$ such that $x \to y$; otherwise, there is no path from z to any vertex in $I(z)$. Clearly, $zxyz$ is a 3-cycle containing z, as shown below.

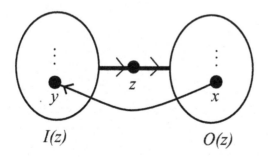

(ii) It is false.

The following tournament T of order 4 is a strong tournament.

Note that the arc xy is not contained in a 3-cycle. So (ii) is false.

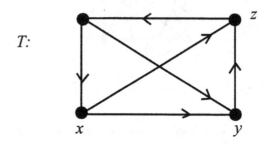

T:

(iii) It is false. In the tournament T shown in (ii), the arc xz is not contained in a Hamiltonian cycle.

(iv) It is true.

Let xy be any arc in T. As T is strong, there exists a $y - x$ path P in T. Clearly, P and xy form a cycle which contains xy.

(v) It is false.

In the tournament T shown in (ii), there is no Hamiltonian path from x to z, nor Hamiltonian path from z to x. $\square$

Books Recommended

(i)   J. M. Aldous and R. J. Wilson, *Graphs and Applications: An Introductory Approach*, New York: Springer, 2000.

(ii) F. Buckley and M. Lewinter, *A Friendly Introduction to Graph Theory*, Prentice Hall, 2003.

(iii) G. Chartrand, *Graphs as Mathematical Models*, Prindle, Weber & Schmidt, 1977.

(iv) G. Chartrand and L. Lesniak, *Graphs and Digraphs*, Chapman and Hall, 1996.

(v)   G. Chartrand and O. R. Oellermann, *Applied and Algorithmic Graph Theory*, New York: McGraw-Hill, 1993.

(vi) G. Chartrand and P. Zhang, *Introduction to Graph Theory*, New York: McGraw-Hill, 2004.

(vii) J. Clark and D. A. Holton, *A First Look at Graph Theory*, World Scientific, 1991.

(viii) J. Gross and J. Yellen, *Graph Theory and Its Applications*, CRC Press, 1999.

(ix) F. Roberts and B. Tesman, *Applied Combinatorics*, 2nd edition, Pearson/Prentice Hall, 2005.

(x)   P. Tannenbaum and R. Arnold, *Excursions in Modern Mathematics*, 5th edition, Prentice Hall, 2004.

(xi) W. D. Wallis, *A Beginner's Guide to Graph Theory*, Birkhäuser, 2006.

(xii) R. J. Wilson and J. J. Watkins, *Graphs: An Introductory Approach*, New York: Wiley, 1990.

Index